Sport 2.0

Sport 2.0

Transforming Sports for a Digital World

Andy Miah

The MIT Press
Cambridge, Massachusetts
London, England

Set in Stone Sans and Stone Serif by Toppan Best-set Premedia Limited. Printed and bound in the United States of America.

Library of Congress Cataloging-in-Publication Data

Names: Miah, Andy, 1975- author.
Title: Sport 2.0 : transforming sports for a digital world / Andy Miah.
Description: Cambridge, MA : The MIT Press, 2017. | Includes bibliographical
    references and index.
Identifiers: LCCN 2016018952 | ISBN 9780262035477 (hardcover : alk. paper)
Subjects: LCSH: Sports–Technological innovations. | Performance technology. |
    Mass media and sports. | Social media. | Communication in sports–Technological
    innovations. | Sports spectators–Effect of technological innovations on. |
    Olympics–History–21st century.
Classification: LCC GV745 .M53 2017 | DDC 796–dc23 LC record available at
    https://lccn.loc.gov/2016018952

10 9 8 7 6 5 4 3 2 1

for @ethanmiahgarcia #play

Contents

Acknowledgments

A number of people have influenced the journey this book has taken. The first is Dennis Hemphill, with whom I found much common ground in the initial stages of writing. Hemphill's work informed my own broader interests in digital technology as not only a mechanism of activism but also a tool for constituting performance spaces and conditions. A second important collaboration during the process of writing was with FACT in Liverpool, particularly through its role in the development of the Abandon Normal Devices festival, which began as a London 2012 Cultural Olympiad project in collaboration with HOME in Manchester and Folly in Cumbria.

My Fellowship at FACT allowed me to engage with a new range of sport stakeholders within the digital art community, many of whom found an entry point into sport through the Olympic Games. Conversations with various people over the period leading up to London 2012 enriched my thoughts about what this book could contribute. I am particularly grateful to Mike Stubbs, Dave Moutrey, Gaby Jenks, John O'Shea, Leon Seth, Heather Corcoran, and Laura Sillars for the thoughtful conversations we have shared over the years about all things digital. Special thanks also go to Debbi Lander, the London 2012 Creative Programmer in England's North West, who was a strong advocate for my work on citizen journalism through the #media2012 project, and who was a champion of digital innovation around the London 2012 Games. It was also a pleasure to work with Drew Hemment and the Emoto 2012 team during the Games, which allowed us to create a unique, artistically informed visualization of London 2012 Twitter data.

Other people who have informed this journey are Emma Rich, Kris Krug, Ana Adi, Beatriz Garcia, Daniel Dayan, Monroe Price, Charlie Beckett, Nick Didlick, Larry Katz, Alexander Zolotarev, and Alex Balfour. Their thoughtful contributions during discussions and correspondence have influenced the book's range of considerations, and their views became constitutive of the research process that underpinned the work along the way.

I am also especially grateful to the many people within the Olympic movement whom have provided support for my work over the years. In particular, I have felt privileged to work with Anthony Edgar, Head of Media Operations at the International Olympic Committee, whose support and generosity have allowed me to develop the research during the period between London 2012 and Rio 2016. I am also grateful to conversations with Mark Adams, Dick Pound, Alex Huot, and Alex Balfour, and Hisham Shehab for their willingness to share perspectives and insights into Olympic matters and the world of sport.

I am grateful also to Alex Lim and Jay Shin at the International E-Sports Federation, and to Patrick Nally. Many of my ideas about the future of the competitive e-sport industry were crystallized through conversations with them at the seventh eSport World Championships in Seoul during 2015.

Some thanks are due to those who have been directly influential to the funding of this work. First, I would like to thank the University of Salford in Manchester for supporting my broad portfolio of research, and to thank my former employer, the University of the West of Scotland, for its investment in my research over the last decade. Additionally, thanks to the British Academy and the Carnegie Trust for the Universities of Scotland, which funded my empirical work during the Olympic Games from Athens 2004 to Vancouver 2010. Finally, I thank Doug Sery for continuing to believe in this project as it developed from one Olympic Games to the next.

These acknowledgments are offered not just to the time we have shared in the development of my data gathering, but also to the time I have spent wondering what they would think about my overarching ambition for the book and my pursuit of imagining this new vision of sport. I hope the book does justice to the one element that unites all the people I have acknowledged above, which is their willingness and desire to transcend boundaries in their respective endeavors.

Introduction

Intelligent exoskeletal devices (data gloves, data suits, robotic prostheses, intelligent second skins, and the like) will both sense gestures and serve as touch output devices by exerting forces and pressures. ... Exercise machines increasingly incorporate computer-controlled motion and force feedback and will eventually become reactive robotic sports partners. ... Today's rudimentary, narrowband video games will evolve into physically engaging telesports ... .
William Mitchell (1995, p. 19)

William Mitchell's vision of future human–computer interactions helped to shape my interest in the relationship between sports and digital technology. His vision of a world where "intelligent exoskeletal devices" augment the range of human functions and the sensory experiences we enjoy resonated with my own views about the direction sports would take—a view that was also influenced by what was happening in biotechnology. In the late 1990s, cyborg researchers were drawing attention to the common ground shared by digital and biological systems, revealing new possibilities for how their integration could permit our experiencing a new kind of corporeal presence. In this context, it was becoming clear how such approaches to being in the world could create new kinds of possibilities in the realm of performance, not just in sports but in music and dance too.

Small changes in established sports also suggested that the structural parameters of sports were not sufficiently robust to accommodate the changing biological capacities of techno-scientific athletes—athletes whose careers, minds, and bodies had been shaped by insights from sport science and technology. For example, the ever-increasing speeds of men's tennis serves generated debate about changing some elements of the sport's physical dimensions—for example, increasing the size of the ball or raising the height of the net. While members of the governing bodies of established sports considered how to modify their games to maintain their integrity, others considered how new kinds of sports, designed for these enhanced humans, may emerge. In the case of the latter, Mitchell imagined a world of "remote arm-wrestling, teleping-pong, virtual skiing and rock-climbing"—a veritable feast of cyborgian experiences made for our growing bionic capacities.

Mitchell's posthuman future coincided with another of my influences, typified in the performance art of Stelarc (Smith 2005). Rarely does one find reference to Stelarc in the sports literature, but his pioneering work in exploring the cyborg interface has relevance for how one imagines the future of sports. Stelarc's exoskeletal machines and digitally immersive devices offered a glimpse into a future in which our movement and thoughts would be mediated by technology—a future in which artificial intelligence converges with robotics and new forms of human agency bring forth new ways of experiencing embodied action. Back in the 1990s, many of these possibilities were realized only in the creative performances of artists such as Stelarc, in the novels of authors such as William Gibson, and the writings of intellectuals such as William Mitchell. Some of the ideas seem crude today; when they were first articulated, however, rapid accomplishments in digital technology were beginning to show how such scenarios could soon be realized. As the new millennium began, the development of digital technology by a new generation of netizens was provoking a shift in how people consumed media, and a population of "prosumers" (Toffler 1970) was beginning to emerge. These new digital communities were more concerned with producing digital media content than with consuming it, and this growing desire to be active rather than passive in our technological culture does much to help explain why these possible futures are so compelling.

As digital devices and sports cultures develop, humanity comes ever closer to an era of virtually constituted sports performances in which the primary medium of participation is not a physical playing field but a digitally mediated space. Consider the recently launched Oculus Rift experience produced by the company Virtually Live, which uses motion-tracking technology to capture the movements of players within a live soccer match. It then translates the data into a computer-generated Oculus experience, allowing the user to feel as though he or she is a spectator within the stadium, sitting in the stands and watching the match in real time. A number of questions pertinent to this book arise from these prospects. For example, how would such conditions change sports experiences, physical activity, and people's sense of what it is to be embodied? How would the technology change the social meaning attributed to sports, the social function of sports, and the way in which sports create participatory communities? Would sports begin to occupy a different place within our social and cultural lives, if our experience of them is played out in virtual realities? Furthermore, what are the consequences of making corporeality a surrogate to a virtual economy, thus creating a physical culture that is defined largely by digital interactions? Would we even make the distinction if the simulation were perfect?

Finding ways to answer these questions—and others that follow from them—is what interests me about the subject of digital sports. This book begins by considering how such technologies challenge how we think about performance, liveness, and the idea of the virtual, then explores how sports are delivering new kinds of experiences through digital technology. Thus, the book first investigates what is understood by a number of concepts that are brought into question by these developments. Specifically, it considers the meanings of "sports," "games," and "play" and how our understanding of them changes when they are situated within a taxonomy of digital leisure practices. It also explores the differences and the similarities between the two primary cultural experiences under discussion: sports culture and digital culture. For instance, how does play within computer culture differ from play within sports? Are there similarities that explain their convergence and that permit one to argue that games occurring within virtual worlds should be afforded the same status as sports? These initial inquiries also outline the range of digital sports subcultures that have roles to play in articulating the rich history of what I call "Sport 2.0," a term that denotes a transition from an analog to a digital way of producing and experiencing sports.

Experiences within virtual worlds have already become inextricable from many other aspects of living. From remote surgical procedures in medicine to managing the global economy, life online is a constitutive element of many societies around the world today. And to varying degrees, participation in a digital economy enables people to traverse other technological divides. For example, in developing countries with limited technological infrastructure or little economic stability, the use of mobile telephones has been a crucial part of the local economy for at least a decade (Plant 2003), and the rise of smart devices is growing especially quickly in such areas. Furthermore, digital products have become a constitutive feature of the creative and cultural industries—which include sports—and they are intimately connected to how licensing, sponsorship, branding, and a host of other creative media practices are monetized. Digital products are also central to strategies for optimizing the commercial potential of any brand.

Perhaps most crucial is the fact that life online occupies the space around our most intimate (private) and most collective (public) experiences. For example, in November 2015, when terrorist attacks by ISIS took place in Paris, one of the most immediate reactions was from the social network Facebook—not just the users, but also the company. Their collective intervention was to encourage their users to change their profile pictures so as to incorporate the colors of the French flag. Overnight, millions of Facebook users' identities became politicized by Facebook's enabling them to take part in an act of solidarity, which changing one's profile picture was designed to convey. Thus, a social-media platform had allowed millions of people to unite around a single gesture, fusing a universal symbol with a unique image—one's photograph—in an act of visible defiance. Of course, these are not simply gestures of solidarity, and later in the book I will consider the complex geopolitical effects of such gestures and how one cannot regard social-media platforms—especially large ones—as simply politically neutral social spaces. Indeed, this aspect of social media raises important considerations for why they often act as editorialized platforms, not just distribution networks for content produced by other editors.

How processes such as those described above are affecting sports experiences, and, more broadly, what this may mean for how we make sense of the role of sports in society, have been largely overlooked. Moreover, while aspects of the subject are pertinent especially to the internal logic and ecosystem of sports, they also speak to wider societal concerns. For instance, chapter 10 discusses how users of social media reacted to the involvement of British Petroleum (BP) in the 2012 London Olympic and Paralympic program. BP was a leading domestic sponsor whose involvement attracted considerable controversy and resistance, most of which was made manifest within digital environments. In this case, the reactions were underpinned by a wider concern about climate change and the use of fossil fuels. The Olympic Games are often subject to similar attacks, criticisms, and even

[handwritten margin note: French flag ISIS]

violence, which make it especially useful to consider them as an indicator
of global social concerns.

Sport 2.0 also provides a way towards re-evaluating the assumptions we
make about digital culture, as may be said of computer games. Despite having enjoyed more than twenty years of life online, computer game cultures
remain a subject of popular controversy. Computer game cultures are still
criticized for their supposedly generating more passive populations, addic-
tive habits, or even violent and anti-social behavior. These allegations are
often directed at specific, prominent examples of digital games, such as
Grand Theft Auto (an action-adventure game set in fictional American cities
where the player's goal is to ascend within an organized crime community).
Yet these allegations make dubious assumptions about what takes place
within such environments; they also tend to treat gaming as a singular
community, when in fact it is diverse. Alternative forms of gaming indicate
how digital participation can inspire extraordinary levels of creativity and
imaginative engagement to the point where "gamification" has become an
ideal means of engagement across a range of sectors. For instance, in recent
years the Wellcome Trust, a large science funding body, launched a program
to gamify one's PhD, and many forms of citizen science programs employ
gaming, as a means of encouraging involvement within such projects.
The alternative depictions of gaming I wish to discuss draw more on its
capacity to nurture politically engaged citizenship and a rise in creative
practice. This capacity becomes even more apparent when the format is
aligned with sports.

In each case, digital technology must be seen as an opportunity for
creative expression, or social engagement, rather than as detracting from
it. Indeed, anxieties about life online may reveal themselves historically
as concerns of a pre-mobile digital age, a period of time in which society
was anxious about change, the erosion of the physical world, and uncertain about how digitally mediated communication was changing how we
relate to one another. As mobile devices transform computers into hand-
held objects and, increasingly, wearable technologies, being online is
looking very different from how it looked twenty years ago. Today, life online
is more akin to the taken-for-granted value we attribute to electricity or other
essential services. Indeed, some societies are beginning to advance the idea
that access to the Internet should be treated as a human right. Understood in
this way, life online and life offline are not obviously qualitatively different.

In this context, this book considers how discourses on digital culture apply to sports. For example, although most of our digital experiences are still mediated through our fingers and hands, does our evaluation of their worth change once these experiences are transformed into activities that demand more physical exertion from us, and, thus, activities in which a clearer relationship between the physical and the virtual experience is apparent? We begin to see glimpses of this in the self-tracking technologies of mobile running applications. Recently the worldwide Pokémon Go craze has generated new conversations about how mobile technology can create new opportunities for physical activity, exploration, and seeing physical places differently. Should it become clear that our evaluation of the digital world changes as a consequence of such trends, our evaluation of the worth of time spent in digital worlds might also change.

This book examines the creative use of emerging digital technology within sports culture in order to reveal the complex ways in which practices such as those mentioned above are changing our views about digital space. Both sports and digital worlds are changing, and a significant part of that change may be attributed to their convergence—digital experiences are becoming more physically enabled through wearable technologies, and sports are becoming more digitized through sharing, big data, and immersive spectator experiences. In this context, I explore how the culture of physicality that surrounds digital life is transforming how we make sense of the desirability of these developments. More precisely, I argue that the advanced use of digital technologies in sports transforms them into new kinds of cultural experiences—experiences that are defined by different values and expectations and which are constituted by new populations of practitioners. In short, digital technology is changing everything about sports culture, including the people taking part, the places where it occurs, and the purposes to which it is put. In turn, the changing culture of sports—marked by the rise of alternative sports—is causing the digital environment to change.

Yet there remains a degree of ambivalence about the value of digital technology within the sports industry, at least as a tool for changing how sports should be played. Critics argue that technology is increasingly dehumanizing the athlete's experience, perhaps even overtaking the human contribution to the results of competitions—a trend that many traditionalists resist. Such arguments are not unique to digital technology; for example, they have been advanced toward a range of technological changes in Formula

One auto racing. Alternatively, the value of digital technology in sport has been questioned over its tendency to replace the role of officials. Thus, in an era of pervasive camera technology, the playing field becomes a living manifestation of Jeremy Bentham's Panopticon in which automated devices govern and regulate the behavior of players as well as ensure that more accurate decisions are made about important results. In such a world, the role of a human referee appears to be compromised and some aspect of sport may be lost as a result.

Despite these concerns about the erosion of human agency in sport as a result of digitally embodied physical activity, there are compelling societal reasons to support such trends. For instance, "Sport 2.0" offers a preview of an era in which sports will no longer be played in environmentally unfriendly places in the physical world, which often become "white elephants" for cities or cause large-scale disruption in resource allocation. Instead, sports will move toward a more sustainable, digital world in which their only environmental burden will be hard-drive space and electricity use. Already one can observe growing support for such new sporting worlds through criticisms about the unsustainability of some sports (such as golf, which is often played in countries with limited water supplies). The rise of the eSport athlete—athletes whose competition is playing computer games—may thus be a prologue to a new era for sport in which such events as virtual golf would thrive. Consequently, this book also investigates the concerns that surround the growth of our life online and the impact of convergence between virtual and non-virtual worlds. In this context, I introduce the notion of *second-wave convergence*, which focuses attention on the sharing of content and on the means of production as a conditioning characteristic of digital culture. In turn, this characteristic is also shaping sports experiences, which become increasingly tied to the utilization of digital technologies.

The broader implications of the increase in digital immersion are already apparent off the playing field. Expectations of participatory digital media have become constitutive of 21st-century citizenship, and the sports spectator's experience is increasingly indicative of this. Furthermore, the debates about a growing "digital divide"—which were prominent in early studies of life online—have shifted in the past ten years, requiring a more nuanced view when explaining what is happening within less digitally developed nations. Although it is difficult to deny that there probably will always be a digital divide, the characteristics of this divide have changed. We now are less concerned about access to technology and more concerned about access to knowledge systems that make participation possible. For example, while one might appeal to the ubiquity of mobile devices or blogging platforms, the proliferation of device technology, operating systems, and open-source solutions requires the end user to quickly adapt to new interfaces that demand ongoing reskilling in order to remain a participant. To some extent, this reflects a shift from Web 1.0 to Web 2.0, but this is just one way of characterizing what has changed.

Despite the challenge associated with democratizing digital technology, there are reasons to be optimistic about the empowering consequences of the shift in the divide which I describe. For instance, the success of Sugata Mitra's "hole in the wall" program to bring computing to areas of considerable poverty, by just leaving computers out in the open for people to use, shows how creative approaches to distributing computing technology can be a gateway to a wider education and even a vehicle of education reform, especially in places with limited infrastructure and public services. People, it would seem, can figure out how to narrow the digital literacy divide themselves, but to do so they need the digital divide to be a thing of the past. The ambitions of Facebook and Google to bring the Internet to the parts of the world that don't yet have it offer a glimpse of the radical transformation the world is about to see.

Overall, this book brings together various aspects of digital technology in sports, covering activities ranging from Olympic competition to computer gaming and remote spectatorship. The Olympic Games, which have always been a showcase for media innovation, provide a way to observe these developments over time. For instance, Olympic Games were the first events to publicly showcase such televisual innovations as slow-motion replay and live satellite broadcasts. The research will show how this innovation is advanced not only by owners of media technology but also, increasingly by users. The Games are also interesting since there are a range of activities that operate *around* the sports competitions, which are gradually connecting the digital worlds of sport and culture. Through new media technologies, the Olympic Games have become an incubator for the use of novel fan experiences, such as the development of urban screens, providing digital public spaces for celebration that now accompany hosting an Olympic Games. These spaces expand the audience experience and give rise to new opportunities for thinking about how audiences encounter sport remotely. Furthermore, they transform how people engage with public space and become a defining component of collective celebration. Additionally, mobile technology has become an integral part of the Olympic Games broadcasting experience, the torch relay (more accurately, a flame relay) playing a leading role via mobile devices. Athletic competitions mirror this use, from the utilization of digital navigation and tracking devices in sailing to the use of Hawkeye technology in tennis umpiring, digital technology is part of the fabric of the elite sports competition.

Thus, the Olympic Games are an arena in which new technologies are implemented and sports take on their role as instantiations of human evolution, both the athletic performance of the participants and the technological grandstanding of their entourage seeking to demonstrate how far humanity has advanced. Making sense of these embedded identities of, especially, elite sports is crucial to understanding their future trajectory and social role. Furthermore, the Olympic Games give rise to remarkable spectator innovation, both in terms of how the Games are viewed remotely and as a vehicle for creating new kinds of spectator experience. This is why the book also focuses on how digital technology is transforming what the Olympic Games mean to their *audience*, which encompasses both sports fans and those who regard the Olympics as an important social movement.

The expansion of digital technology offered through my characterization of "Sport 2.0" also offers more opportunities for the Olympic brand to reach more people, though I endeavor also to reveal where these opportunities may give rise to social concern. For example, the large urban screens that occupy cities during Olympics constantly displaying the Olympic messages may also be seen as a form of indoctrination into celebrating an event which many people do not believe to be simply politically neutral as Olympic family would want to argue. This raises questions about the omnipotence of mega-events and their capacity to undermine expressions of citizenship, exemplified by *State of Exception*, Jason O'Hara's film about preparations for the 2016 Rio de Janeiro Summer Games. In this context, digital resistance becomes an indirect consequence of the Olympic industry and a key mechanism through which citizens can challenge these impositions. In this context, the book documents and scrutinizes the transformations of sports that is occurring through digital technology, while considering the ideological questions it provokes about how people make sense of themselves and their societies in a post-analog age. Thus, I will explain how these two trajectories—from the individual to the societal—are intimately intertwined in the context of sport and digital technology. Furthermore, I consider what this means for the future of performance and participation in one of the most universal of cultural practices: sport.

The book is divided into three parts and ten chapters, which take the reader through the central questions arising from the development of "Sport 2.0." Part I focuses on how digital technology is changing sport experiences from a wide range of perspectives, while also providing some of the book's philosophical underpinning. Chapter 1 considers the different cultures of sport, digital technology, and the Olympics. It explores how much further there is still to go before one can talk about a global digital culture that has become inextricable from all aspects of our lives. It also discusses how sports cultures have begun to change and, in particular, become subservient to media change, and what this will mean for how various systems of governance develop their approach to culture. This inquiry leads to questioning what it is that makes sports experiences distinct and meaningful—in short, their social function and value—a theme that is taken up later in the book. This chapter also explores the societal *justification* for sports, so as to understand how digital technology challenges or responds to these interests. Finally, through analyzing Olympic culture (perhaps the most prominent example of an ideology-driven sports-related program), chapter 1 considers how the Olympic movement has become a central driver in shaping the values of sports culture and business and what it will need to do in the future to retain this place in the sports system.

Chapter 2 sets out the theoretical dimensions of the book. It begins by discussing the current conceptual understandings of virtual reality, computer culture and sport. It identifies the book's major themes, arguments, problems and possibilities, including reference to the major, pertinent philosophical notions of games and sport, integrating the seminal work of Bernard Suits and Brian Sutton-Smith, and how discussions about digital culture and sport relate to them. This is where the book's main questions are formulated and where the pivotal concept of *unreality* is explored to identify areas of common ground in theoretical work on sports and digital culture. In particular, I consider the way in which a certain kind of sense of the physical world arises within digital space, often through the *design interface* that mediate our experiences. Keyboards, mice, and gesture technology are crucial determinants of the language used to discuss the value of computer games and sports as mechanisms of embodied action, limited by design, but capable of being disrupted by new forms of interaction design.

A latent question within this discussion concerns the prospect of sports simulation and whether the optimal interface will be that which permits the seamless simulation of a sports event to such an extent that there would no longer be a need for real-world or off-line sport spaces. This question is not unique to the sport case, but is discussed more explicitly toward the end of the book as a fundamental consequence of the inquiry and a very real possible future for sport. The chapter also discusses how examples of life online require us to reconsider what we acknowledge as real or meaningful in human experience and how this evaluation is contextualized within a set of ideological assumptions about the nature of virtual realities. It introduces the idea of second-wave convergence to explain how sociotechnical changes within such practices as sports give rise to new evaluations of life online. Furthermore, it discusses how our nostalgia for analog lives is particularly apparent within practices such as sports, which are constituted by a presumed notion of what embodied action and corporeality should entail. In pursuing this argument, the chapter also questions how one defines embodiment and considers, in particular, how wearable technology is challenging the view that digital identities are separable from our analog lives.

Part II draws together these analyses and focuses on the broad ways in which digital technologies have affected elite and amateur sports experiences, along with considering how they have re-constituted the spectator's proximity to the action. Chapter 3 examines how digital technologies are affecting the *elite* athlete's experience of sport, but also the other individuals around the athlete, such as officials. The first section of the chapter focuses on performance technology and discusses various examples of elite sports simulations and the emerging digitization of movement, which underpins new sport technologies. While it cautions against claiming that digital technology is radically extending an athlete's potential, it explores changing processes in athletic experience, notably in training, which arise from digital devices and systems. Moreover, it discusses how knowledge arising from digitization is shaping an athlete's experience of sport. The chapter also argues for the virtualization of physicality within a range of sport forms, both elite and non-elite.

Chapter 4 focuses on how the *amateur* athletic experience is being modified by digital technology and how this requires us to re-evaluate computer culture. It focuses particularly on game-based experiences—both the development of computer games and the use of devices that create game-like experiences for an amateur athlete. Moreover, it demonstrates how digital gaming has become an integral part of the amateur sporting experience and suggests why this is important. For example, it discusses the specific context of physical education and how various pedagogic assumptions about computing technology that are apparent within education may be challenged as a result of these technological transformations. The *embodied intelligence* of virtual gaming can provide a further mechanism through which to teach sports and to promote new forms of socialization. The chapter also considers specific examples of physically active games—often called *serious games*, or what Taylor (2012) describes as "embodied play"—which are emerging as alternative forms of sports activity. These examples support the claim that gaming technology is becoming more sport-like, thus challenging the assumption that computer game playing necessarily leads to a more sedentary lifestyle. The rise of the e-sports gaming industry is indicative of this at an elite level, but so too is the growth of digitally enabled communities of physical activity. This chapter also sets up the argument for framing the book through the idea of "Sport 2.0," denoting an emerging sports community that is beginning to occupy the place of traditional sports and which has the potential to overtake it in numerous ways.

Chapter 5 moves toward analyzing the spectator's experience of digital technology in sport. Amateur athletes are part of the spectator community and there are clear ways in which amateurs are connected with elite athletes through spectatorship. However, I also propose that spectating is changing through the development of digital interactive experiences, such as urban screens, TV on demand, mobile technology, and social media. To this end, this chapter focuses on such transformations to the spectator experience, toward what may be called, *remote participation*—a concept that alludes to Sherry Turkle's (2011) insights into digital culture. The chapter also asks whether the concept of spectator still makes sense in the context of immersive viewing experiences where the witness is brought into the space of the activity, rather than simply occupying a third person perspective. The most recent example of such a trend is the GoPro and VISLINK head camera broadcasting within the NHL, pioneered in January 2015.

Part III shifts the discussion away from the culture of sport practice toward the specific context of sports production and consumption, by discussing the Olympic Games, as an exemplar of digital media innovation in sports. The chapters in this part will appeal particularly to scholars of media change, as it documents the rise of citizen journalism and alternative news production practices that occur around sports events. Many of the arguments presented here are informed by my empirical research around the Olympic Games from Sydney 2000 to Rio de Janeiro 2016.

Chapter 6 provides an historical examination of how the Olympic movement has courted relationships that have allowed it to innovate and operate at the cutting edge of media production. It also explores how the expansion of the mega-event media industries have led to greater exclusivism over reporting privileges that has narrowed the lens through which reporting takes place—even if the coverage volume has increased. It also explores the detail of how the Olympic Games stimulate discussions about media change around the world.

Chapter 7 focuses on the emergence of new journalist communities at the Olympic Games, which articulate how its media community has grown. It argues that the expansion of the Olympic "fringe" journalist community results from the exclusive arrangements that surround sports reporting, but also the growing expansion of large sports events to become more like cultural festivals, which attract the interests of non-sports reporters. In so doing, the chapter charts the rise of the non-accredited media center and its strategic role for Olympic hosts, made possible by the extended means of reporting via digital technologies. While the chapter urges caution in claiming that this expansion reveals a trajectory toward greater media freedom at the Games, it does identify how media expansion is changing the way that traditional media organizations operate, provoking a democratization of media expertise and the re-professionalization of journalism.

Chapter 8 focuses on how the rise of social media has transformed media events. First, it considers the characteristics of the Web 2.0 era and what influences this has had on mediating sports cultures. Second, it considers how the Olympic industry has organized its response to this new communication architecture, providing guidance to its community. Next, it explores how social media may threaten the financial base of the Games, considering how to monetize Olympic social-media content. Subsequent sections in this chapter consider the risks of open media, the expansion of the user experience by digital technology and the parallels between open-source volunteers and the Olympic volunteer ethos. In so doing, the chapter articulates a vision for digital culture that is born out of the values of social media, as an ideological force that coheres with the Olympic vision and with a broad perspective on the potential contribution of sports in society.

Chapter 9 provides a detailed analysis of the social-media interventions surrounding the London 2012 Olympic and Paralympic Games, widely discussed as the first social-media Olympics.[1] It examines how social-media platforms were instrumental in generating news content during these Games—not just distributors of the news of others—while also discussing how the organizing committee, stakeholders, and audiences contributed to generating the record breaking volume of social-media content that came out around these Games.

Chapter 10—the culmination of the book's trajectory from the individual to the societal, which began in part I—focuses on how digital culture is shaping citizenship, by creating alternative channels of communication, activity and expression, which demand our occupation. Whereas chapter 9 explored the celebration of the Olympic Games through social media, chapter 10 considers resistance to these expectations, by discussing examples of Olympic protest and antagonistic reactions online. These forms of alternative or disruptive media narratives are, I argue, a necessary component of the Olympic industry and a crucial means through which the Olympic ideal can assert itself as a movement, rather than just a sports event.

The conclusion consolidates the case for how sports are becoming increasingly digitized practices and why "Sport 2.0" is a necessary way of imagining sport's future. It also expands our consideration of digital life toward biological configurations, building on earlier chapters to describe how digital technology is transforming the athlete's *biology* and how this changes the conditions of future sporting encounters. It discusses the implications of these ideas, which encompass the need to remove sports from their physical worlds and to relocate them in digital space. Furthermore, it acknowledges how the interface between the biological and digital worlds will transform sports and other physical cultures in the future, for instance, through increasingly intelligent prosthetic devices.

While writing the book, I often asked myself "What relevance will a book about digital technology have in such a rapidly changing world?" Partial answers can be found in an understanding of the history of such changes and in an acknowledgment of the importance of resisting the pressure to elevate the importance of real-time occurrences over historical occurrences or future trends. However, the relevance of any book about technological change is more crucially found in recognizing a series of trends that suggest what sport's future will look like. In this sense, Sport 2.0 may be seen as a new chapter in this history, as we begin to weave together physical experience and the digital world.

I   The Field of Play

Two concepts that persistently recur in debates about innovation in sport and digital culture are *games* and *play:* Whether one is trying to ascertain whether a new technology will disrupt the intended test of a game or whether new social-media platforms can be gamified to create new kinds of experience, these terms have an extensive body of research behind them.

The chapters in part I consider what forms of human endeavor occur around such concepts to reveal their importance in debates about Sport 2.0. Each term is underpinned by a theoretical context that informs the arguments presented in chapter 1 and the shifts that are taking place within the worlds of digital culture and sport. Each concept is also informed by a body of literature that has shaped an understanding of sport and digital innovation, but which also has been crucial in its characterizing the relationship between work and leisure, a distinction which finds itself in turmoil within present times, where many people find themselves both always at work and always at play, or never completely doing either. In short, inquiries into games and play reveal fundamental characteristics of our human condition, which are pivotal in explaining the power of digital technology to change the circumstances of our lives.

The notion of *game playing* brings together the present debate about sports and virtual worlds. Indeed, both our participation in sports and our participation in virtual worlds may be treated as forms of game playing. Consider the quotidian description of sports as "only games" and its corollary, that they are more than games. The first description is used to console someone who is (overly) disappointed with losing a game, or is directed toward someone who seems to be taking a game too seriously and failing to identify the other values that participation in it involves. Labeling something "only a game" appeals to the idea that we should not place too much stock in the results of sports competitions, since they are not really what matters in life. Yet in the past thirty years sociologists and cultural studies scholars of sport have challenged this view, arguing more on behalf of the latter—that sports games are perhaps even more important than life outside of them. Thus, scholars argue that sports, because of their important cultural and political role but also because of their effects on the lives of participants or of spectators, should be understood as indistinguishable from life off the playing field. For instance, it is not inconceivable that someone's life as a sports fan has greater currency in who she considers herself to be as a person than does her working life, her career, or even her family. This interpretation of sport's social function reveals the changing role of leisure experiences in the twentieth century, but also draws attention to the alienation arising from the world of work, which is well documented by scholars as being a product of a post-industrial society.

To understand the complexity of these impositions on our chosen fields, it is necessary to investigate the language through which normative judgments are made about different states of reality, since these decisions determine the value and the importance we attribute to such pursuits. In other words, if what we do when playing a game is not seen to connect with our broader personal, social, economic, and political life, then it is easily dismissed as trivial or inconsequential in the broader scheme of things. Although it is easy to think of examples of human endeavor that fail to fulfill a directly instrumental role in society, the challenge of games is that they deliberately eschew the wider world by setting up special rules that constitute a different kind of reality: one that is goal directed and is designed to control the conditions of a certain test. The function of games is precisely in providing alternative spaces in which people have the capacity to explore who they are and what they can do. Like virtual worlds, sports operate within a kind of cyberspace (to use a term from the early years of Web studies). Yet relegating sports and life online to the status of merely a game simplifies their role in constituting our social, economic, cultural, and political world. Sport, as a narrow instance of physical movement and mental acuity, may thus be interpreted as a crucial but not an exclusive route toward comprehending our own embodiment. Were it not for sports (and dancing, and playing music, and exercising), we would fail to know ourselves as physical beings. Alternatively, digital worlds help us change our sense of boundaries and the limits of our identity and potential.

The problems arising from such playful explorations of cyberidentity are discussed in detail in chapter 2, where I offer a response in the context of sport and virtual worlds.

Some of the earliest work in Web studies and in sport studies focused on our need to make sense of games and game playing as an overarching construct that shapes the experiences of each, and indeed there is considerable common ground here. For the former, computers lend themselves to rules and rule keeping, requiring set protocols, heuristics, language, and conventions that create the conditions of our interface and our experience. Similarly, sports rely on rule making and adherence to achieve a particular kind of experience. In part, this explains why sports lend themselves to controversy, since breaking rules betrays the tacit agreement made between competitors to maintain those rules. This self-evident truth precedes the moral debate about the significance of rule transgressions. In this case, the value of the event is constituted by maintaining the rules and is, by implication, compromised when they are broken. Thus, the breaking of rules matters not only from a moral perspective, but also from a purely functional perspective in which the sense of an experience collapses when rules are broken. Various bodies of literature have theorized these relationships, and scholars from a range of disciplines have made important contributions to our understanding of what games are and why they matter. (See, e.g., Suits 1967 and Suits 1978.) One of the primary contributions of this book is to place these bodies of literature alongside each other and examine how they help us make sense of the challenges arising from Sport 2.0—sports that are played out in digital space.

As was intimated above, play and game playing are also prominent concepts within debates about how work and leisure are related, which is also a crucial theme in today's digital culture as the boundaries between the two begin to blur. Personalized media experiences through mobile devices and a greater integration of digital systems within work and leisure practices force into sharp relief the inadequacy of claims that the virtual can be seen as a distinct space in our lived reality, separate from the non-virtual. Yet adjusting to these changes takes time, and there remains resistance to their separation within such cultural practices as sport. In this sense, the challenge from digital worlds may be seen as fundamental and total, given their capacity to undermine the category of the physical. After all, what may give special value to human performance is its undeniable embodied solitude, isolated from the increasingly pervasive digital noise of present-day life. If that singular condition is digitized, the quality of the experience may be jeopardized, simply because technology gets in the way of the physical experience and becomes an obstacle to sensorial immersion.

The "field of play" for this book is shaped by these underlying conceptual debates about play and games, sport and digitality. If there is a single premise underlying the book, it is that present times demonstrate how these ideas and the cultural practices in which they are played out are converging to such an extent that it no longer makes sense to think of our lives—or ourselves—as ever being offline. Physical and virtual worlds have become indistinguishable, but not only in the sense that we spend a lot of our time using mobile phones, computers, or tablets. Rather, the virtual is becoming physically constituted through the integration and occupation of digital technologies into all aspects of our lives, and this trajectory extends beyond sport. This means that it is necessary to examine other practices in which these processes are apparent. In sport, the complete integration of the virtual and the physical is becoming comprehensively apparent, and it has the potential to dramatically transform the culture of other practices, giving rise to a new kind of digital physical culture.

1 Games-Based Culture

Understanding Digital Culture

In the early 1990s, one of the expectations for the Internet was that it would reconstitute—or perhaps more fully realize—Marshall McLuhan's (1964) "global village" by creating new patterns of communication, post-industrial innovation, and new ways of organizing societies. Furthermore, the Internet was thought to have the potential to redistribute power away from a few large media organizations and toward a wider population of media producers and owners. In turn, these circumstances would reconstitute power structures within societies and, it was hoped, make the world a better place. Not all technological innovations are designed with such lofty expectations. There may be some expectations that an invention should improve something, but only rarely is an invention expected to improve the political conditions of the state or the entire world. Closely related to this desire was the expectation that transforming the means of media production would give rise to a more participatory society in which more people would have means of expressing their views, which would then have influence in the socio-political sphere.

In parallel, many of the discussions about Internet policy and investment strategies were—and remain—focused on how digital technology would revolutionize industries and institutions, bringing about a change in how they conduct their business. That sentiment was echoed in May 2016 when the British government called for global action against corruption through the use of digital technologies to make governments more transparent and to enable grassroots communities to blow the whistle on corruption. On this view, social change would occur as a result of transformations of trade and commerce, creating new, disruptive forces that would reconfigure how money is made, how businesses are created, and what kinds of services are sought. A good example of this is the mobile-application economy, which has created new forms of revenue around new services such as photo-editing applications, health tracking, and social networks.

More than thirty years of scholarship in Web studies has shown that some of these expectations neglected to consider how new injustices would also arise from new digital configurations. Indeed, there has been no radical shift in the processes governing media ownership. It remains the case that a small number of providers control the channels of communication for the majority of people. Moreover, even though today's new media elites include giant organizations such as Google, Facebook, and Twitter, which have provided new means of expression, they are vulnerable to the same kinds of criticisms as older media giants. Perhaps the only difference between the old and the new media entities—insofar as cultural change is concerned—is that the latter are content *hosts* rather than content *generators*, which shifts their political role considerably. However, even this distinction may become harder to substantiate as monetized, promoted content begins to proliferate within such platforms. It is even harder to justify this claim about the value-neutral function of social-media platforms as more effective tools of monetizing data aggregation create personalized media exposures for users that enable the owners of the providers to filter what audiences discover. The consequence of such processes is that the platform becomes a content generator, even if it is a process of curation rather than creation.

Despite the worthy aspirations of Howard Rheingold (1993) and other early cyberlibertarians, the range of people who are engaged with developing digital worlds—as opposed to just consuming them—was much narrower than experts foresaw when the Internet became a popular communication device. Indeed, the Internet's rise was accompanied by discussions about an emerging "digital divide." That term referred to the fact that many people did not yet have access to the Internet or to computers at all and thus had no ability to engage with the so-called Fifth Estate. These observations brought caution to the utopian views of the Internet in an era in which both scholars and strategists proclaimed its potential to transform society, as they did until the burst of the first "dot-com bubble" on March 10, 2000. But although access to technology was a central concern of the Internet's critics, the underlying liberal concern was that failure to access the Internet meant failure to gain access to civil society, through which one could participate in democratic forms of public negotiation.

Despite considerable growth in online access (for example, Africa's online population increased by 2,000 percent from 2000 to 2010), concerns about division and access remain an enduring currency of the Internet era. As of 2015 only 9.8 percent of the population of Africa was using the Internet (Internet World Stats 2015). Moreover, discussions about its potential to liberate people in ways that may not be possible to achieve by other means are recurrent themes within Web studies and new-media research. The 2009 election in Iran is a case in point: The emancipatory aspirations for the Internet were re-asserted once more, in this case generating attention around the use of the micro-blogging platform Twitter. Twitter was discussed in the media as having been an influential platform for getting information out to the rest of the world from a country where the global media had only limited access to information. Grossman (2009) and Morozov (2009) even assert that the US Department of State requested that Twitter postpone scheduled maintenance at the time of the election to lessen the likelihood that communication with the outside world would be disrupted. For this reason, Streberny and Khiabany (2010) claim that "Twitter" became shorthand for a broader revolution in political blogging in Iran. The aspiration and the wider cultural shift toward online political campaigning reinforce this claim.

Although evidence suggests that the most influential tweets about the situation in Iran were generated outside of that country, the important point is that new communication configurations can become influential platforms for engaging other communication channels, such as television or print publishing. For some time now, what takes place within social media has become worthy of news coverage in other media channels and sports news is fast becoming a part of this process. Thus, despite the fact that the use of Twitter within Iran at the time was relatively limited, its reach extended far beyond the significance of the platform thanks to the manner in which influential Twitter users shared content and the mass media's coverage of such activity. Indeed, one might even hypothesize that social-media environments such as Facebook and Twitter obtain their primary positioning due to the uptake of a relatively small community of early adopters. Consider, for instance, the Twitter user Stephen Fry, a British comedian, writer, and public intellectual. In 2008, Fry was among the top ten Twitter users worldwide, which included major media providers such as CNN. Fry's alliance with Apple's iTunes, his prominence in British debates about our digital future, and even his prominence in the public eye undoubtedly contributed to Twitter's popularization. Though mediatization has always been a process in which members of an intellectual elite drive information toward the masses, this is also what makes a focus on digital technology and sport so fascinating, as much of it emerges from mass utilization, particularly by those who are not engaged in the wider discussions about digital democracy.

Twenty years after the initial discussions about the digital divide, many circumstances of the digital world have changed. The divide has narrowed. More people have access to the Internet in some form, whether through digital television, home computing, public libraries, or mobile devices. In China there has been a remarkable increase in access to the Internet. In 2008, China, with more than 400 million mobile phone users, overtook the United States as the country with the largest Internet user base. By 2013, this figure had grown significantly; use of phones for access to the Internet was also growing quickly. By 2009 there were approximately 230 million mobile Web users; by 2013 there were more than 460 million; by the end of 2014 there were 557 million (Reisinger 2015); by mid 2015 there were 668 million. This growth is not occurring only in urban environments, either. Half of the new Internet users in China are in rural areas (BBC 2012). Consequently, the divide between different countries has also been reconfigured, if not narrowed. In this respect, it is less crucial that the technology access gap has narrowed and more important to understand how populations that previously had no access to the Internet are now able to function in some meaningful way, by acquiring the appropriate literacy skills.

This does not negate the fact that there is an ongoing need to engage with digital divide management. Although home computers may have become cheaper and easier to use, and even publishing online may have become something that relatively unskilled people can do (particularly via social media), the early adopters still retain the ability to reap the most benefits from digital technology, since they are best placed to exploit its value. Moreover, as practices of online publishing and broadcasting have been de-professionalized via freeware, there has emerged a greater fluctuation and change in the software knowledge required to utilize such environments. For example, some years ago Web designers relied heavily on one or two specialist design packages to create personal homepages. In contrast, the past five years have seen a proliferation of dynamic server platforms—Blogger, Wordpress, and Tumblr, to name a few—which require comparably few skills and much less financial investment to utilize. Indeed, many of these platforms are free to use, but each requires an ongoing re-skilling by users, which many people may not have the capacity to undertake. Whatever is currently the particular platform of choice, the capabilities that are now possible include live streaming and broadcasting of video content, multi-camera automated editing, and even the creation of virtual-reality content. Whereas once these capacities would have required considerable funds, they are now offered by free or moderately priced platforms online and on mobile devices. Perhaps the biggest challenge that societies face today in regard to digital media concerns people's ability to understand how these new platforms work, since new environments are created daily.

Thus, even though the digital technology divide has diminished, there is a perpetual digital literacy divide. In some ways that divide is more alarming than just inequalities of access to technology, since it is a divide that is not easily diminished by just providing infrastructure or technology. Rather, narrowing this gap relies on a continuing ability to develop knowledge-based skills that can allow as many people as possible to enjoy the range of ways in which digital technology is progressing and providing new opportunities. Of course, this divide is also partly about money. Consider *World of Warcraft*, the most lucrative online game ever. In many ways it epitomizes innovation in online gaming, signaling the growth of networked, mobile gaming. Yet to participate in *World of Warcraft* and in other innovating computer cultures one must have paid a subscription fee and must have a broadband connection whose speed accommodates the demands of the game, which itself has a cost attached to it. Though these costs don't exceed those of other computer games (approximately $200 per year), not all will be able to afford them. The same is true of the more recent game *Minecraft*, around which a vast industry network has emerged, though its one-time cost of about $30 is much more affordable. Nevertheless, if society is to attend to the problems presented by the digital divide it will have to address questions of digital literacy.

Despite these concerns about present-day digital culture, computers have influenced all aspects of developed societies. Even when computers are physically absent from our lived encounters with social systems, they underpin our daily practices—driving cars, riding trains, using credit cards, and so on. Computers have established and support global channels of distribution and economic infrastructures, and this configuration of influence is neatly captured in Castell's (1996) notion of the "network society" which envelops all of us. In commerce, politics, education, and leisure, information and communication technologies have altered—and will continue to alter—human capacities, and have reproduced complex modeling algorithms that help explain biological processes, mechanical systems, and the social world. Thus, computer-generated models and simulations are crucial mechanisms that underpin such areas as architecture, design, air-flight training and management, medical surgery, transport systems, ergonomics, as well as in tourism and education. Moreover, a computer-based innovation in any one of these areas often finds application in another, reinforcing the value of the networked society.

The complexity of our computing histories continues to grow and is, on one level, a history of how innovation intertwines over time. For instance, gamers now enjoy a rich and diverse history of computer gaming, from the mechanical pinball machines of the 1970s to arcade environments in which such activities as virtual motor racing and horse racing take place. It also encompasses the more widely experienced form of home computing, which offer a range of digitally generated simulations of such activities as motor racing, football, skiing, boxing, and basketball. These examples of the innovation history of digital culture convey its richness, the nuances of which are rarely detailed when computer games are scrutinized in the mass media. Rather, computer culture is often treated in a singular manner. Yet one can make only limited claims about these different stories as evidencing linear progression within digital development. Although computer processing power increases, graphical realism gets better, and connectivity speeds increase, these improvements are not necessarily what all computer communities desire from their gaming experiences. And in the 21st century's first decade, computing nostalgia emerged as a distinct category of popular computer culture.

Consider the game consoles of the late 1970s and the early 1980s, such as the ZX Spectrum and the Atari consoles. Games from such developers have been reanimated by the development of emulator devices—often flash-based consoles that allow users to replay such games within new computer operating systems, usually via a Web browser, but also through plug-and-play television apparatus and mobile devices. Game emulators and game genres that are built on relatively crude graphical, auditory and narrative structures have become popular as free, streaming games on the Internet, through such websites as miniclip.com. These technologies allow those who enjoyed computer games from twenty years ago to revisit those experiences, though the space in which they encounter the games has changed dramatically. The wireless gesture controlled object has replaced the fixed position, dynamic joystick of the 1980s, and the high-definition, surround-sound quality of the experience has replaced the low-graphic, monotone musical experience that characterized early games.

Making sense of these trajectories is crucial when attempting to provide some analysis of what might be the social worth of innovation within computer culture. If one compares the birth of the computer games industry to the birth of cinema, then the industry is still in an era analogous to that of the period of silent film. Even the digital spaces in which gaming takes place are changing. Now, some of the largest social-media networks allow the integration of online, free gaming experiences, which often consist of simple platform games or retro games. This republishing of early console gaming experiences within social-media platforms creates new cultures of computing and new modes through which digital culture is enriched, embellished, and reaffirmed. They also create new spaces of leisure activity— functioning as nostalgic devices, rather than exemplars of the cutting edge of digital technology. Such games embody different values within gaming culture compared with the development of high-tech game consoles, which trade on the values of graphical realism, narrative complexity, and Turing-like interactivity. Today, mobile gaming fills the void that we encounter in between places, a defining characteristic of modern urban lives, where commuter time and mobility become spaces for digital and social catch up.

These new gaming spaces reflect a different kind of cutting edge in digital technology—mashing up a spectrum emulator with Facebook is a gaming feat that extends innovation within gaming culture. This act of mashing up re-invents the experience of retro gaming—and provides a new chapter in a game's history, which challenge our assumptions about what gaming communities want in terms of graphical improvements. In these cases, the primary value is of re-living previous gaming lives. Online mobile gaming, in its many forms, has opened up new ways of re-defining gaming culture, which even includes the revival of board gaming. Consider the prominence within Facebook of Scrabulous, a game that attained enormous popularity in 2008 and, in so doing, became the subject of one of Facebook's first trademark battles when the proprietors of the board game Scrabble acted promptly to assert their rights under the Digital Millennium Copyright Act and stop the unlicensed proliferation of their game on this growing social-media platform (Ahmed 2008). The result was a removal of Scrabulous from Facebook later that year for North American users (Goel 2009), to be then re-launched as the same game but called Lexulous, under a ruling made by the Delhi High Court in India.

Understanding Sports Culture

These stories of game playing online lead neatly into debates about cultures of sport participation. Indeed, an equally prominent feature of 21st-century culture is that of sports participation and spectatorship, which have become increasingly big businesses in the past thirty years. Though the economic downturn of 2008 placed some sports in jeopardy as major sponsors withdrew from various competitions and clubs, sports have grown vastly in absolute economic terms year upon year, often as a result of specific transformations in the exploitation of sports media content, but also through transformations to how sports contracts are assembled. For example, the *Bosman* ruling in European soccer prevented clubs from "demanding 'transfer fees' for out-of-contract players who move from one club to the next" (McArdle 2000). Its effects were numerous and included transfers across countries in Europe, though one of the most important changes was to a player's ability to negotiate new contracts, which dramatically increased the amount of money players could expect to earn. This brought massive transformations to how soccer was financed and a vast change to the industry's worth. Soccer was no longer just a serious game; now it was a big business attracting large financial investments.

A central part of this was the growing financial value of athletic celebrity, whereby the licensing of an athlete's name through merchandise also changed dramatically over the same period. Today, a top-level athlete in a high-profile sport may license anything from cologne to insurance. An athlete's in-competition appearance may also be branded to such an extent that nearly no part of the televised body is lacking a logo of some kind. Indeed, during the lead-up to the 2012 London Games, the American athlete Nick Symmonds launched a campaign on the popular auction-based website eBay selling advertising space on his body. The winning bid would have the right to tattoo his or her Twitter account name on Symmonds' arm for display during the 2012 Games. The description for the lot read as follows:

> Here is a very unique opportunity to advertise your Twitter name on an Olympian during the 2012 season. I was a member of the 2008 USA Olympic Track and Field Team and am currently ranked #1 in the United States and #6 in the world at 800 meters. I will wear a temporary tattoo of the Twitter name of the auction winner in every single competition I run in this year, 2012, on my left shoulder.

This was the first instance in history in which the branding of an athlete by a sponsor involved physically marking an athlete's body, though it should be noted that the proposal was for only a temporary tattoo. The proposal to display a Twitter name rather than a logo also raises novel questions about the significance of logos in a world in which usernames—rather than a visible design—become the primary brand identifiers. The winner of the lot was—perhaps not surprisingly—a marketing company. Its Twitter handle was @HansonDodge. The company paid $11,000 for the privilege and subsequently leveraged its investment by making some promotional videos with Symmonds (Elliott 2012). Although it was never likely that the International Association of Athletics Federations would allow Symmonds to display the tattoo during competition, the act conveys an innovative and provocative attempt to challenge the current model of athletic sponsorship, which favors only the most successful athletes. Indeed, the entire system of Olympic sponsorship rules was questioned by athletes leading up to the 2012 London Games through the #Rule40 and #WeDemandChange Twitter campaigns. The campaigns drew attention to the IOC's rule that prohibits athletes from exploiting their image for commercial gain during Games time (Rogers 2012) and it remains a live concern ahead of Rio 2016. Ahead of Rio, the campaign website for Rule 40, makes the following assertions; they are reproduced here as they appear on the site, without capitals:

only half of american track and field athletes who are ranked in the top 10 in their event earn more than $15,000 a year from the sport.

a majority of athletes live way below the poverty line. they are forced to scrape together a barely-livable income made up of prize money, sponsor contracts, grants and part-time jobs.

unlike athletes in many other countries, american olympians receive no direct support from the federal government.

the us olympic committee offers health insurance and stipends to only a limited
number of competitors.

Similar examples are easily found, and each reveals a situation in which
sport became increasingly immersed in exclusive transnational corporate
business interests. A related example of the pervasiveness of branding is
provided by host-city contracts for the Olympic Games, which indirectly
stipulate that while the Games are in progress the host city must ensure
there is no ambush marketing. This translates into all public billboard space
being occupied with sponsors' posters or being left blank, thus transforming the entire into an all-consuming advertisement. The complete occupation of public and private space by sponsors—symbolized by Symmonds'
intervention—may be one of the most alarming consequences of digitally
commercialized sports, since it overshadows all other values that sport may
seek to elevate. These examples of digital innovation invite us to consider
whether sport's future is at all desirable, or whether these circumstances
offer a glimpse into a future where all aspects of human labor are translated
into only financial value.

A further facet of sport's social role is its contribution to the promotion
of healthy lifestyles, which engages a key governmental agenda. Some evidence of sport's impact on health is found in studies of how, around the
world, sport and exercise have gradually become intimately connected to
medical and health research objectives. University departments of health
science, sport, and exercise research are increasingly expected to produce
work that can have an impact on such circumstances, particularly where
they can shed light on troublesome aspects of behavior, such as eating dis-
orders (Rich and Evans 2005), or novel techniques that can help to alleviate
the economic healthcare burden, as may be said of genomics (Miah 2004).
Using sport to promote a healthy lifestyle has been on the agendas of governments for many decades—perhaps even since before 1896, when the
modern Olympic Games began. Though this alignment may not always be
apparent from the relatively low numbers of hours attributed to physical
activity in schools, or from the implementation of policy, or even from the
lived reality of participation in sports (which can be very unhealthy), sport
is often used as a political device for the promotion of health.

In elite sport, athletic success is also intimately tied to national aspirations and identity. Sports results are often discussed as being symbolic

of national strength, an interpretation of sporting societies that has been a strong characteristic of national value for decades—perhaps since 1936, when, it was widely thought, Hitler used the Berlin Games to spread Nazi propaganda. The first modern Olympic torch relay was an integral part of the Berlin program, and today the torch relay is one of the most pivotal parts of how the Olympic narrative is orchestrated, which helps build local public support for the Olympic program. Despite the ideologically abhorrent context of the 1936 Games, the principles that developed around Berlin's Olympic propaganda have found utility in subsequent Olympic Games host cities, which have sought to use the Games as a platform for place marketing and advertising. Moreover, the development of sponsorship and television rights revenue packages led to an advertising-based sports culture. More recent examples of such showcasing of national strength are the 2008 Summer Games, held in Beijing. Those Games can be seen as the last to conform to the mega-events model, as a new narrative of environmental sustainability and responsible event hosting emerges in their wake, with more modest aspirations—and budgets. Indeed, the suspension of the ambitious stadium design for the 2020 Tokyo Games is indicative of a growing apathy for mega-events. As Tokyo's stadium budget began to spiral out of control, the loss of public support meant that an alternative approach would have to be found.

The licensing of digital products, such as computer games, has led to the sports genre becoming the most financially lucrative of all genres of computer games. The reasons for this are simple to explain: Sports games became mechanisms for advertising, and their licenses are renewed each year, as new competitions take place. With each new license, new contracts are attached for advertising. That provides a reliable—and competitive—means of investment. For instance, the FIFA World Cup computer game is reissued by EA Sports every four years and is accompanied by annual updates and licenses related to the UEFA European Football Championship and the UEFA Champions league. By 2016, it had sold over 100 million units worldwide. Similarly, the Olympic Games computer game is re-issued for every Summer and every Winter Olympics.

Despite the obvious commercial imperatives that influence the development of sports, sports organizations have also always promoted the idea that they perform a broad social function as well, either as a mechanism for generating positive role models for young people or as a route toward opportunities for social learning and building international relationships. Perhaps the most prominent sports organization with such ambitions is the International Olympic Committee (IOC), which is partly why I focus on the Olympic Games in the final part of this book. I will discuss later whether the IOC fulfills its aspirations to make the world a better place—or how it could further its achievements—but it does more than other elite sports organizations by imposing such obligations upon itself. Some of these values are pertinent to the present debate, but there are many aspects of this work that extend beyond the remit of this book. For example, the IOC program Olympic Solidarity is a mechanism through which revenue generated by the Olympic Games is redistributed to provide funds for athletes from developing nations to travel to the Olympics. Alternatively, the Olympic Truce is an initiative that involves the IOC's working with the United Nations to promote the cessation of conflicts between nations during the Games. While scholars have criticized the achievements of these programs, they tend not to exist in any form outside of the Olympic infrastructure, and so it is reasonable at least to credit the Olympic industry for attending to these matters at all, even if more could be done to ensure they succeed.

As further demonstration of these broad social aspirations, the IOC has always required that a condition for winning the right to broadcast the Olympic Games involves showing how a bidding broadcast can reach the largest free-to-view television audience. Such a self-imposed obligation may be jeopardized in a difficult economic climate or in a declining market for television advertising. For example, in 2009, the subscription sports television provider Setanta Sports went into administration, and the IOC began talks to, potentially, offer the rights to a pay-per-view provider for the Olympic Games, though it indicated that the major competitions would be required to be free to view. And in 2011 the European Broadcast Union—of which most national European broadcasters are members—lost its right to broadcast the Games, also suggesting an era of pay-per-view Olympic television on the horizon. In any case, a commitment to reaching the widest audience has been a central part of the IOC's Charter for some years, and the IOC argues that pay-per-view still allows for this, though with a much reduced number of television hours. Furthermore, the recent €1.3b-billion multi-platform deal with Eurosport—whose parent firm is Discovery

Communications—signals another new era whereby all rights (across all platforms) are brought together under one contract (IOC 2015b).

Thus, the aforementioned commercialization of sport may be seen as peripheral to its primary social value. After all, despite the omnipresence of sponsorship within elite sports, what distinguishes elite sport as a cultural endeavor is its capacity to create competitions in which elite performers make history by transcending the presumed limits of human capabilities, evidenced by the breaking of records or by the performance of extraordinary physical feats. While this theory of sport does not fully describe the value generated by an elite athlete's performance, it is a necessary component of what distinguishes elite sport from recreational competition. Were it not for the potential of experiencing extraordinary human achievements, elite sports would not hold the value they enjoy.

These aspects of sport culture reveal how it is intimately connected to broad present-day values and issues, particularly those that concern our lives within digital space which are fast becoming embodied, corporeal experiences through such innovations as wearable devices that track our movements using sensors. Their dual trajectories signal sport's and computing's next era of innovation, via the creation of physically demanding and digitally constituted practices. The implications of their transformation are far reaching when one considers that, traditionally, sports and digital culture have been seen as polar opposites. Indeed, the rise of computer culture is often seen as directly correlated to the erosion of physical activity.

Though many governments around the world still treat sport as a form of cultural practice, they are distinct from other cultural industries in numerous ways. There are different theoretical approaches to articulating these differences, from the essentialist debates about the nature of sport or culture to the broadly empirical claim that what distinguishes sport from other aspects of cultural activity is the mere fact that sport occupies more space in newspapers than any other cultural endeavor. This simple observation nevertheless calls for us to acknowledge the fact that an initial means of comparison between various cultural endeavors is how they are reported by the media or, more broadly, how they enter the socioeconomic, political, and cultural spheres.

One should always be wary of claims that a particular cultural enterprise is unique, but sporting culture enjoys characteristics that allow it a distinct place in people's daily lives, through their involvement as either participants or as fans. Perhaps it is the allure of witnessing historic achievements that gives elite sports such value, but it may be the fact that sport spectatorism often develops in fans during childhood, it becomes an affiliation that we grow up with. Indeed, some theories of sport analogize the experience to a form of religious or spiritual enlightenment—a claim also advanced for other extraordinary creative expressions, such as musical performance and dance. Thus, it may be unwise to claim that what distinguishes sport's value is completely different from other forms of cultural practice. Yet sport's widespread popularity operates on two levels. First, it relies on the mediatization of celebrity as a modern-day form of hero worship and adulation. Second, it locates a commitment to a team or the pursuit of competition as a distinct community of interest, around which a range of other social experiences may take place (collective viewing, shared identity through clothing, chanting, and so on). Whether or not one is persuaded by this tribal explanation of why sports matter to people, their prominence in many cultures around the world is powerful on various terms: financial, political, cultural, social, and so on. Perhaps it is the idea that sports rely on a universally shared language that explains their value. Sports are made possible only by the participants having a shared purpose within an agreed and non-negotiable set of rules. If one considers how participation in sport is enabled by such rules functioning across many levels of governance, national and international, the power of sport to leverage its interests are considerable.

Yet understanding sports requires further examination. Sports business today has reached such heights of regulated activity that nearly every part of an athlete's body is governed by a precise and detailed set of expectations for generating sponsorship revenue and licensing. Moreover, beyond branding an athlete's body, sponsors' logos are now present in the physical structure of the field or court, as well as in the architecture of the arena. Consider how branding technology within sports arenas has changed over the years. Whereas in the past physical billboards were erected at the borders of stadiums, now there are geometrically designed in-pitch graphics and widgets within televisions screens, targeted to optimize recognition to viewing audiences through television.

There are very few exceptions to this overwhelming presence of branding within sports, and in this regard the Olympic Games are both an exemplar of such exploitation and a guardian of principles that protect the neutral space of competition. For instance, sponsors' banners may be hung only on the *exterior* of an Olympic stadium, not inside it. Whether such a principle is sustainable remains to be seen, and here again digital technology may be instrumental to an impending transformation. After all, as mobile devices become increasingly important to the broadcasting—or narrowcasting (meaning the personalization of mediated content)—experience, the first mechanism through which sponsors will enter the stadium in a major, organized fashion will be via content delivered by augmented reality within mobile devices. In this respect, the integration of the live spectator experience with *second screen* viewing is the point at which the capacity to limit the exposure of spectators to sponsors becomes unsustainable.

Sport-based computer games are no exception to these processes, and the presence of advertisements within the architecture of computer games has become an important mechanism of their achieving prominence. Indeed, sports computer games should be seen primarily as a mode through which advertisers engage and develop their target audience and through which brand loyalty is furthered. These economic parameters of the industry should be in the foreground when one is considering what underpins sports gaming cultures. A cynic may view gaming culture as simply a vehicle for supplying loyal followers to a wider economy, rather than an end in itself.

Understanding Olympic Culture

As was explained in the introduction, the study of digital technology and sport is well articulated through the historical development of the Olympic Games. The Olympics provide a useful example through which one may track the range of ways in which industries have operated around elite sports practices. Over the years, media technology has been pioneered at the Olympic Games, and this offers a way to analyze how sports culture has changed progressively in the past hundred years. Furthermore, because the Olympic organization is distinct in its being underpinned by an ideology, most notably articulated within the Olympic Charter, the Olympic infrastructure provides the essential connection between sport and culture that I wish to explore in order to address the ways in which their relationship has become constitutive of many values that operate within society.

For example, the value held in the symbol of an Olympic gold medal may be attributed to the status that the Olympic Games have accrued as a celebration of human excellence in its broadest sense. Indeed, these values pervade the Olympic infrastructure, thus ensuring that the latest innovations are often first implemented within the Games, as opposed to other sporting spectacles. Moreover, this promotion of "excellence" as a brand value operates within the non-sporting dimensions of the Olympic experience in its entirety, which encompasses the desire to showcase world-class art within the Cultural Olympiad. Indeed, this is evident at an Olympic Games in the form of the sponsors' pavilions, which have become a showcase for emerging technologies—such as Lenovo computers or Samsung mobile phones. Though there has been no mandate to require such sponsors to use their Games-time presentation as a showcase for the future, recent Games demonstrate the elevation of these values, as do the press materials produced by the sponsors during the Games. Here again, there is an overlap between the manner in which both sport and digital technology aligns with the metaphor of progress—athletes as examples of our evolutionary achievements and digital technologies as evidence of our mastery of the science of craft.

The range of organizations that constitute the Olympic Family, as it is often called, can be difficult to summarize in a single formal structure. As the legal guardian of the Olympic Movement, the International Olympic Committee holds the proprietary rights to (and obligation to protect) the Olympic brand. To this extent, it must be seen as the "supreme authority" on all Olympic matters. At present it consists of 98 members and 206 National Olympic Committees, and it affords recognition to International Sports Federations, which vie for this accolade. Constitutionally, the IOC Charter outlines the broad aspirations of the Olympic Movement, which encompass explicit humanitarian and social goals. These goals were an integral part of the original revival of the Olympic Games in the modern era, which came about under the direction of Baron Pierre de Coubertin in 1894. Coubertin's vision for an international revival of the Games occurred in a period when a number of international movements were developing. Indeed, there had already been two unsuccessful attempts to revive the Games. It is widely thought that the global dimension of Coubertin's proposition secured its success, though another reason may have been

Coubertin's aspiration for the Games to address social malaise within France, where young people were disenchanted with social conditions.

The literature on the revival of the Games is extensive, but most relevant here is the fact that the modern Games are characterized by an explicit set of philosophical and social values that compel the organizing body to address a range of non-sporting agendas that resonate with the aspirations of digital innovation. The work of such aspirations is evident in a number of the IOC's activities, perhaps no more strongly than in the revival of the Olympic Truce in 1992. The mechanism through which the IOC's president makes this appeal is the United Nations general assembly. The IOC has cultivated this relationship with the UN for many years; for instance, in 2009 the IOC was granted Observer status at the UN.[1]

Another of these goals is a concern for promoting sustainable development via responsible environmental planning. The environment has become an issue of considerable importance for the IOC in the past twenty years. In part because the Olympic Games have a significant environmental impact on a host city, they also leave a very large carbon footprint. The IOC now emphasizes sustainability in its charter, and each host city is expected to minimizes the environmental damage wrought by the Olympic Games. The targets for doing so have often been criticized for imposing unreasonable artificial goals on cities, which have their own governmental targets to meet. The centrality of environmental concerns marks a significant alteration to how the Olympic Movement positions itself with respect to international agendas. Indeed, its working relationship with the UN Environmental Programme (UNEP) is a further indication of this.

The Olympic Charter's high ideals have often met harsh realities, which bring into question the IOC's ability to preserve the Olympic Games as a neutral territory, free from politics and abundant with humanitarian aspirations. Arguably, the IOC's alignment with the United Nations is one example of its having compromised its political neutrality. Nevertheless, moments in Olympic history reveal how the Games often come to symbolize more about society than just sporting excellence. For example, the 1968 Mexico Games became a moment for demonstrating support African American civil rights when the US athletes Tommie Smith and Juan Carlos gave "Black Power salutes" when receiving their medals. At Munich in 1972, the attack on the athletes' village in which Palestinians took Israeli athletes hostage left an indelible mark on Olympic history. Many more such examples—many of which are less well known—demonstrate how the Olympic Games have been a platform on which political stands have been made. When the Olympic flame relay was on its international course in advance of the 2008 Beijing Games, few could have predicted the degree of protest and anti-protest that would ensue, ultimately leading to the cessation of the international torch relay in future Games. For better or worse, these moments have become rich parts of Olympic history, symbolizing the significance of the Games as a global media event. Indeed, the example of the pre-Beijing torch relays a rich story of mass media versus citizen media. In the coverage of the relay, the official route was changed by the Organizing Committee various times, and members of the media entourage who relied on the host were unable to track the progress of the torch. However, in various places, citizen journalists—that is, citizens with Internet-enabled cameras—were able to broadcast what was taking place in real time. Such occurrences reveal how journalism is not just a practice defined by individual journalists or professional bodies, but also a complex series of relationship among journalists, editors, media organizations, and audiences, all of them enabled by digital technology. The third part of this book will focus on how digital technology has become an integral part of Games-time innovation, but not solely from the perspective of technological change. What is distinct about new media at the Olympics is the cultural transformation that has taken place around communication; this significant societal shift is the focus of part III.

In sum, digital culture, sport culture, and the Olympic Games constitute the context in which this book advances ideas about Sport 2.0, the point at which sport became inextricable from digital technologies. In so doing, the book aspires to tell stories of their intersections, which provide an indication of the directions that each will take in its cultivation of Sport 2.0 and a new way of experiencing technology and our own physicality.

2 Real-World Games

A virtual reality comes into existence when a group of people experience a simulation as if it were real.
Randy Walser (1991, p. 57)

One of the common themes of research into both sport and digital culture is the attention that has been given to their *realness*. It is often said that sports operate outside of the real world and that they are essentially non-serious, playful pursuits made possible by their gratuitous logic (Morgan 1994). This logic is constituted by a set of rules that condition a specific activity and distinguish it from others. In turn, these rules prescribe a set of sub-optimal efficiencies that bring into existence the peculiar practices of sports and the means by which goals may be pursued (Suits 1967). For example, baseball involves a batter swinging a uniquely sculptured object at a relatively small, hard ball that has been thrown at high speed from a distance of 60 feet and 6 inches. Hitting the ball would be much easier if either the ball or the bat were a little bigger, or perhaps made of different materials, or if the pitcher were to throw more gently. However, the optimal test of baseball players—it is thought—is achieved by the unique balance of these constitutive elements. There is nothing that dictates their precise relation to one another—the ball could be bigger or could be pitched more slowly, and the bat could be easier to wield. Yet the present-day physical limitations of players require that the sport's equipment be created to these dimensions. That is not to say that they are fixed; rather, it is to say that the perception of what sports are testing and the manner in which those tests are carried out changes. And the components of baseball have indeed changed. For instance, the distance between pitcher and hitter has been changed over the years as pitchers have become more competent.

Thus, the object of the game of baseball is brought about by the reasonably precise configuration of the activity, which is carefully balanced to test specific kinds of human capabilities. While each individual component of this composition may be relatively arbitrary, their collective assembly coheres around a very specific set of abilities that players consider worthy of testing themselves against. For example, if the distance from the pitcher to the batter were much greater, pitchers would not be able to pitch as accurately. Alternatively, if that distance were too short, batters' ability to react to pitches would be diminished. Baseball players accept these configured limitations in order to make possible the specific test within the game, and the game relies on their acceptance. The same is true of all sports: Players must accept the creation of imposed limitations to permit the enjoyment of certain kinds of tests of ability.

It may also be said that sports are unreal to the degree that they operate outside of society's norms and structures. The abandonment of these norms creates circumstances in which athletes are allowed to commit acts of violence upon one another without fear of legal reprisal, as in the case of combat sports. These protected zones of social activity allow unusual forms of behavior to occur and to be celebrated by an otherwise law-abiding community.

These observations are not undermined by also recognizing that what takes place in sports has consequences in the real world. For instance, participation in sports is often a means toward greater socialization among individuals who otherwise would not have such opportunities. Furthermore, the history of sports is replete with examples in which the mere existence of sports has led to tangible social or political change or has enabled people to make symbolic gestures. For example, at the 2000 Sydney Olympic Games, athletes from North Korea and athletes from South Korea were able to compete under a common flag for the first time in history, despite remaining politically separate. Sometimes it is *only* in the simulated, symbolic world of the sports field that such gestures are possible.

Though one might presume that these two forms of unrealness are contradictory, they are, in fact, perfectly compatible. Sports are possible only because of the acceptance of their lusory (unreal) elements, and yet they remain highly consequential in the wider social and political sphere, affecting the lives of millions of people in profound ways. A similar thesis may be advanced in regard to digital worlds: Life within them may be described as unreal or as a form of virtual existence that occurs separate from our lives outside of such environments. Indeed, virtual reality is a foundational notion within these debates, and it deserves special consideration to allow closer analysis of how it is tied to my interpretation of sports practices.

In the 1980s, the virtual reality of cyberspace was described by William Gibson as a "consensual hallucination"—a world brought into being by "lines of light ranged in the nonspace of the mind, clusters and constellations of data [l]ike city lights receding." Today's vision of virtual reality is somewhat less poetic, grounded in the lived experience of virtual worlds, but it remains intimately connected to the distinction between the words *virtual* and *real*. For many people today, life online is as familiar to their modes of communication and daily life as time spent in offline spaces. Indeed, our experiences of life in offline and online worlds often occur simultaneously. People move through virtual realities while walking through physical worlds, as when one is walking while texting, talking, or navigating the Internet. Indeed, increasingly digital games are designed to exploit this duality, utilizing augmented reality to make gaming more immersive, as for Pokémon Go. Consequently, the division between the virtual and the real requires greater scrutiny today, in order to come to terms with what is understood as unreal or real within digital spaces.

In the early 1990s, Howard Rheingold argued that the concept of virtual reality failed to adequately address the real-world consequences of spending time in cyberspace. What are we to make of these propositions when coming to terms with how digital technology operates today? Paul Virilio, Bruce Sterling, Arthur and Marilouise Kroker, and even (to some extent) Jean Baudrillard balked at the idea of a virtual reality that could be discussed as oppositional to some other more tangible reality. Instead, they defended the realness of the virtual, Virilio treating it as a substitution and Baudrillard as a simulation. Despite their differences, both cases treat life in digital worlds as unequivocally meaningful and certainly consequential. So how should we regard the relationship between online and offline worlds if they are increasingly one and the same?

In this chapter, I utilize the word *unreality* to discuss the common ground between digital and sporting virtualities and to explore how they occupy distinct but similar spaces within society. Thus, I undertake the historical and philosophical work of making sense of these spheres of unreality, while appealing to the idea that their *unreal* status provides only a partial insight to their value. The main point is to investigate the connotations of unreality within a wider politics of the public sphere. In this context, unreality becomes an area of concern only when a prior sense of reality is threatened. Understood in this way, the relative realness of digital and sporting virtualities is irrelevant to the value we attribute to each of them, since it is principally their juxtaposition that creates the tension. In other words, when unreality threatens to become a dominant mode of existence (as may be said increasingly of life online), people become anxious about the erosion of life offline.

Such conditions generate a kind of "virtual anxiety" (Kember 1998) as demonstrated by the neuroscientist Susan Greenfield, who often criticizes life online fearing its consequences. Greenfield's underlying rhetoric is suspicious of life in virtual worlds, and she has dismissed Twitter and the wider reliance on computers as detrimental to a richer, more intelligent life that could be led by elevating experiences in offline worlds. Greenfield's anxieties lead me to consider how the computer game—as a distinct cultural form—has been isolated as the locus of social concerns about computer culture and about a decline in physical, creative culture that allegedly follows from its growth. I have deep reservations about the legitimacy of that perspective. I am especially keen to draw on examples from exergaming, or serious gaming, as substantive challenges to this naive critique of gaming cultures.

Although subsequent chapters within the book will assess how computer games transform our expectations of digital culture and sports practices, this chapter emphasizes how they are converging practices with similar internal logics. This is a crucial position to accept in my broader thesis, as the critique of life online falls foul of the fact that life offline—the supposed standard to which virtual realities are judged—is rapidly becoming constituted by virtual systems. From our reliance on contactless payments—or even the Bitcoin[1] economy—to the vast digital systems that underpin the transitioning of goods around the world, life offline is quickly becoming a historical concept as the principle of experiencing everything everywhere comes to define what can be expected from the next era of digital connectivity.

Unreality Bytes

The unreality thesis of digital worlds underpins the additional claim that people would have more valuable, richer lives if only they were to spend less time within such environments. In the mid 1990s, Bruce Sterling wrote of such concerns as being fueled by an anxiety that life in virtual worlds would lead to neglect of life offline:

The good news is that I can chat with distant strangers. The bad news is that while I'm on the Internet, I'm not chatting to my next door neighbour. I'm not going to any neighbourhood rallies, I'm not throwing parties for local friends, I'm not baby-sitting other people's kids. It may be that I'm not even talking to my own children, who are off in the living room being raised by Nintendo. (Sterling 1997, p. 29)

Indeed, there is a great deal of literature detailing the detrimental societal consequences of spending too much time online as a kind of addiction. Moreover, this "addiction" is often used to explain many of society's ills (Funk 2001; Griffiths 2000). The same is also sometimes said about participation in sport, which for many committed enthusiasts may become all consuming, detracting from other forms of socializing and even leading to isolated modes of existence and unhealthy obsessions with physical training. Furthermore, schools' reducing the amount of time allotted to physical education reinforces the idea that sport is less important than more serious activities, such as mathematics or science. At the professional level, the obsessive behavior associated with becoming competitive at an elite level may also lead to high-risk behaviors, such as experimenting with drugs to improve performance, or to corruption via fixing. In this sense, the virtual world and sports arenas are often treated pejoratively, as lesser worlds—not inspirational Second Lives, but second-class lives, opposed to primary, more important modes of existence, where a more wholesome form of social interaction takes place. At best, they are seen as recreative experiences that—to adopt the civilizing thesis of Elias and Dunning 1986, which claims that sport's principle function is to operate in service of life outside of sport—support the working life.

How should one make sense of these pejorative and reductive presumptions about the unreal status of the digital and sports worlds? Now that many people spend more time communicating with others via technological devices than they spend communicating face to face, does it betray technological progress to berate these circumstances or resist them? After all, people quite comfortably inhabit both worlds without much difficulty most of the time. Users of cell phones can still value the moments of contact and immediate connectivity they provide while reviling behavior such as talking loudly on a train or using a phone on an airplane. Sadie Plant's (2003) early work on mobile-phone culture attests to these contradictions. Her informants talk both of feeling that "there's something missing" (p. 64) if they are without their mobile phones. Equally, it is apparent that the device can be "a source of tension, disagreement, and antipathy" (p. 33).

As the use of mobile devices continues to increase year by year, these contradictions are becoming ever more apparent, though cultures also adapt. For instance, for some time it might have been considered impolite for someone to let a cell phone ring during a meeting. Today, at many public meetings, attendees are no longer asked to switch off their mobile devices; indeed, at some meetings attendees are asked to keep them turned on so as to be able to share content using social media. This is not to profess that there is an absence of etiquette in mobile culture, or that all uses of devices in all settings are free from any expectations to disconnect. Rather, it is simply to highlight how conventions around technology change, and to point out that what is considered undesirable or even offensive may change after a period of mass adoption and adjustment.

The concepts of reality and unreality are useful devices for thinking about the relationship between digital culture and sports. Indeed, I am unwilling to relinquish the concept of unreality completely, despite acknowledging that the term raises questions about what takes place within alternative realities. We ought, I think, to pursue unrealities, as much as we pursue reality in our lives. Our absence from the conventions of real-world spaces is crucial to our imagination and our ability to cultivate new ideas. Yet there is also an ontological distinction between these forms of human existence, and the import of this claim becomes apparent when one considers examples of transgressive behaviors within unreal worlds. For example, if an athlete abuses an umpire, assaults a spectator, or commits some other kind of moral transgression of the norms of the practice, the unreality of the practice become a matter of interest to the outside world. Infamous examples of moments when the unreality of sport became a real matter of wider social concern include Eric Cantona's flying assault on a Crystal Palace fan in 1995 and Mike Tyson's biting the ear of Evander

Holyfield in 1997. More recently, Oscar Pistorius' complaints at the end of the 200-meter race in the 2012 London Paralympic Games reveal the expectations people have of athletes to maintain the suspension of their disbelief throughout their victories and their losses. Such transgressive acts—or even just our knowledge that such transgression is possible—gives sports their special social value and makes them an enduring and compelling form of spectacle.

Of course, the naive unreality interpretation of sports clashes with the reality of how it can affect the world in important and consequential ways. In simple economic terms, a nation may invest money in grassroots community sports or in its elite athletic population and, in so doing, may dramatically affect how sport is valued and experienced. Moreover, there are ways in which what happens in sport competitions bears on political relationships between nations. Indeed, sports policies and participation often bring about forms of social change—in South Africa, for example, rugby was utilized as a vehicle for unification. Sports can also provide moments of manifestation and activism that shed light on important political issues. This is further indication of how sports transcend the playing fields and occupy other realities. Moments of this transcendence from unreality to reality demonstrate how sport has become a very serious business indeed.

There are also many stories that explain how the experience of becoming an athlete can have a dramatic bearing on an individual's sense of place in the world, but also how the world responds to certain issues that are played out in politics. Consider how the Olympic Games have provided a mechanism for promoting gender equality in recent years, or how individual Olympic athletes have become important symbolic motifs for specific ideas. The Australian indigenous athlete Cathy Freeman became a symbol of indigenous peoples' rights when she lit the Olympic cauldron—the most privileged non-competition role at an Olympiad—at the 2000 Sydney Games. Her act, and the organizers' choice of her to perform that role, must be explained in the context of years of protest about the lack of respect shown toward the indigenous people of the land on which the Olympic Park—Homebush Bay—was constructed.[2] Alternatively, Muhammad Ali's lighting of the Atlanta 1996 Olympic cauldron had particular historical significance in that it was situated in a part of America where he had experienced racism in the 1960s, even after having previously won an Olympic medal. Though it is often said that such moments are simply tokenistic and do not really address the way in which deep social change occurs, the world would not know as much as it does about racism without these historical sporting incidents, which gave special symbolic currency to social campaigns and which may have been particularly important in enabling populations to unify around an idea about the need for social change. Each instance reflects ways in which the unreality of sports, or the theater of performance, transcends its supposedly artificial environment. The argument that sports are simply game playing, or that they are separate from the real world in which more serious events take place, becomes untenable in such moments, when it quickly becomes apparent that these symbolic acts are among the most important parts of our real-world histories.

Though there can be no doubt that elite sports are often very serious play spaces, their ideology nevertheless relies on the playfulness of sport, in a similar way to how one might discuss other kinds of play encountered through the practices of music, dance, or acting. Thus, the athlete inhabits a social role that symbolizes various primary social virtues—excellence, dedication, struggle, and so on—that imbue their achievements with meaning. However, outside of the specific sports environment these values may hold little functional meaning themselves, and the practice of elite sport may be seen to compromise other social goods. In this respect, one may claim that the primary contribution made by athletes is to enrich our sense of values, which is why so much is at stake when an athlete is found to be cheating. Through their endeavors, athletes enrich a certain notion of what it is to be human and to pursue excellence.

Historicizing Virtual Worlds

As was suggested earlier, a cornerstone of debates about the state of the real in digital worlds has been the attention to virtual reality, the history of which requires further unpacking before one applies it to sport or to present-day digital culture. Hemphill (1995) notes that virtual reality isn't a novelty in present-day society. More familiar (or, perhaps, less obvious) examples of its presence are "documents, phonographs, radios and television" (ibid., p. 56), which each allow a certain kind of escape into alternate worlds. However, virtual realities now are approaching a change in kind rather than degree; at least, the degree of change in virtual-reality technology is becoming increasingly profound. For example, consider how holographic technology has begun to transform our sense of being in the world—our perspectives on the necessity of travel or the nature of being somewhere at all. In 2015, Microsoft launched a product called HoloLens, which it describes as the "first fully untethered, holographic computer, enabling high-definition holograms to integrate with your world." Recent developments in 3D screen experiences are altering people's expectations of what filmmaking entails and what may be possible to express through that art form. Virtual reality is interesting in the present analysis for a number of reasons, though first I wish to consider its symbolic importance.

Ideas about what it may be like to experience a virtual reality—or simply to engage with the idea that it can be distinguished from another reality— have been present in literary forms for centuries. One might even talk of literature as a mode of virtual reality in which the function of narrative is to draw us into other worlds. Explicit engagement with the idea that we might be living in a simulation, or what that might be like, are common themes in stories, such as *Alice's Adventures in Wonderland* and *The Lion, the Witch, and the Wardrobe*, that compel us to think about the possibility of inhabiting alternative worlds. Popular fiction has also frequently devised ways of imagining virtual realities. The 1999 film *The Matrix* presents a world that is indistinguishable from the world we inhabit, inviting us to consider whether or not we would be able to know it if we were living in a simulation (Bostrom 2003). The 2009 film *Avatar* depicts a world in which it is possible for a disembodied mind to inhabit another being and, in so doing, extends the consideration of virtual realities to what it might be like to live cognitively and remotely inside another being. Perhaps the ultimate simulation film of recent times is *Her* (2014), which asks us to consider what it would be like to have a relationship with an artificially intelligent computer operating system that has intellectual and emotional intelligence superior to those of humans. These possibilities of experiencing life in "the singularity" are fascinating because they raise questions about our individual uniqueness, but also because they permit freedom to experience life beyond the single one that we enjoy.

Popular fictions such as those cited above are rooted in ideas that have been played out in other literary and scientific spheres. Thus, one may treat the pursuit of virtual worlds as an attempt to recreate the physical world as convincingly as possible, with various modeling techniques—from the development of figurative drawing to computer aided design—inextricable from the means available to us to imagine and subsequently inhabit such spaces. In the more radical propositions, the prospects of creating virtual worlds are often intimately connected to stories about the rise of machines as intellectual competitors with humans. In this way, debates about life in a digitally mediated virtual world lead invariably to imagining a state in which humanity may, one day, succeed in creating replicant beings— autonomous robots that could replace humanity, as in the 1982 film *Blade Runner*.

In this context, the possibility of virtual worlds is a mesmerizing subject to consider, as it is cloaked in a series of dystopian possibilities that invite response because they may soon be realized if we do not intervene. It is important to acknowledge this when seeking to separate the mythology of virtual worlds from the realities of present-day technology, or even near-future technological applications, since there are considerable differences. This is not simply a reality check; it is a recognition that a wider range of technological changes may result in *biotechnological* transformations. Thus, various altered human states also compel us to imagine what life would be like if lived in a different cognitive and corporeal dimension, whether through drug-induced highs or through extreme embodied experiences such as simulation of zero gravity or any number of "extreme sports" encounters. Indeed, Blascovich and Bailenson (2011) describe drug use as a route through which one may seek out life in a virtual reality. These more familiar human experiences are intimately connected to how one can begin to make sense of what life in a virtual reality would be like and what utility it might have.

Virtual realities envelop our worlds in various ways. Even an automobile could be described as a form of virtual-reality environment in which our experience of faster-than-human velocity occurs in a simulated space, protected by an absorbing shell that shields us from the wind and keeps us dry and warm. Such virtual realities are already strong indicators of what it might be like to live in a virtual world. The fact that we do not think of travel in planes, trains, and automobiles as forms of virtual realities has more to do with the way in which technological systems become seamless modes of existence than with the inappropriateness of our claiming them as modes of virtual existence.

Indeed, the design principles of technological development now aim to promote experiences that minimize the perception of a transition from offline to online reality. The most effective designs are those that make technology absent from the encounter, or what Peterson (2007, p. 79) describes as "mundane cyborg practice." Good examples of this include the "rape in cyberspace" documented by MacKinnon (1997): In a text-based gaming world called LambdaMOO, one member of the community (the violator) took control of another person's persona and proceeded to violate the character in the presence of all other users. It is easy to trivialize this example, by arguing that it was just a group of people playing a game, and suggest that what happened to the character didn't really happen to the person. However, this kind of environment blurs fantasy and reality. People are there with different motivations, some playing out fantasies and others meeting with friends and having "real" conversations. The action mattered to people within its community and this is the basis for which it became a matter of moral concern. Thus, to argue that cyberspace is a manufactured, artificial, and unreal environment raises a question about whether anything at all that takes place in cyberspace is real, but the more obvious conclusion is to state that it all matters.

MacKinnon (1997) argued that, although it can be questioned whether the LambdaMOO incident did constitute rape, there was no doubt that "the current iteration of rape as constructed in LambdaMOO poses serious, real consequences for users of virtual reality." Indeed, the incident was widely publicized and was said to be the event that turned a "database into a society" (J. Dibbell, 1993, cited in MacKinnon 1997) and was widely cited as a case in which a virtual world suddenly achieved a state of mundaneness such that what took place within it was seen as every bit as serious as if it had happened offline. This is not to suggest that rape in any form is mundane; it is, rather, to suggest that the subjection of this virtual world to wider social conventions and expectations is what allows it to make the transition into meaningful, real, and banal space, rather than it being a place of simple fantasy. In this respect, becoming mundane may be considered an aspirational value of all technologies, as it denotes the status of ubiquitous acceptance as part of the offline world. These aspirations are built into the design interface of new technologies, though it is often hard to achieve. A good example of this design aspiration is found in the case of Google Glass. Despite its withdrawal from production in January 2015, only two years after its limited beta release, Google Glass was always promoted as a device that would take "technology out of the way," removing the need for an obstructive interface and creating an experience that was more seamlessly connected to our sensory intuitions. Yet it was among the most intrusive personal digital devices ever to have existed.

This perspective on the distinction between real and virtual is made more conspicuous in Annette Markham's 1998 book *Life Online*, in which the distinction is reviewed in the context of cyberspace research. As Markham notes, "most authors presuppose a particular understanding of the term real ... and often contrast real—when talking about computer technologies—to the term virtual" (p. 117). Yet this distinction is often not fully problematized; in fact digital-reality interactions can be as real as sensory-realities, since they are not fictitious realities. As Hayles (1999) writes, "merely communicating by email or participating in a text-based MUD (multi-user dungeon) already problematizes thinking of the body as a self-evident physicality" (p. 17). Additionally, as Turkle writes (1995, p. 180), "now, in postmodern times, multiple identities are no longer so much at the margins of things. Many more people experience identity as a set of roles that can be mixed and matched."

There should be no mistaking the degree to which mundane technologies are, historically, remarkable and radically transformative. After all, what would it have meant for a person living in the 1830s to have seen something like today's iPhone, 3D film, or holographic projections? Such innovations would have undoubtedly appeared to be magic, as Stivers (2001) might put it. Yet it would be flippant to imagine that such technologies would have changed societies dramatically then. Indeed, one of the most infuriating aspects of trying to retrofit technology into the past is the failure to consider the impact of any technology outside of the range of socio-technical systems that permit it to function and, thus, be meaningful. A laptop computer magically sent back in time to the 1830s without a software or energy infrastructure to enable its use would be of little use or interest. Considering technology outside of its specific social context rarely makes sense. This reminds us that considering the socio-technical dimensions of technology—that is, thinking of technology as systems rather than artifacts—is critical to understanding how we should regard it (Bijker 1995). In this sense, as Pargman (2000) notes, "computer games are never 'just games,'" but are part of wider social configurations.

Central to the idea of virtual reality are the notions of *immersion* and *interactivity*. Different virtual systems entail varying degrees of one or both of these elements, but they are each always present to some degree. *Immersion* generally refers to the ease with which one inhabits a virtual world and the degree to which it compels us to feel as though we are in a genuine other world. Depending on the nature and sophistication of the technology, the experience may feel more or less believable, natural, and seamless in relation to so-called real-time movements, sounds, and shapes. Immersion can be facilitated with the use of three-dimensional optic displays, binaural sound systems, and movement-tracking devices. Additional feedback can come from optic-fiber data gloves or body suits that replace bodily sensations with computer-generated ones so that virtual body contact points have sensory correlates in the user. *Interactivity* refers to the degree of control the user has to affect what happens within the virtual world and its capacity to affect the user. How one negotiates these environments may depend on simple keyboard, mouse, or joystick maneuvering or may include kinesthetic/haptic feedback—that is, control of action through special gloves or hydraulically controlled motion platforms.

Despite more than thirty years of development, digital virtual realities have yet to satisfy the science-fiction writers who first imagined them. Most have failed to generate completely immersive experience, although since the launch of Oculus Rift, Samsung Gear VR, HTC Vive, Microsoft Hololens, Playstation VR, and Google Cardboard there has been renewed interest in these possibilities. Yet to expect VR to deliver near-world experiences any time soon is to miss the point of their promise. In fact, they are never likely to exist in the way that was imagined. The "digital delirium" of the 1990s was never a delirium brought about either by science fiction or by cyberlibertarians such as Howard Rheingold. Instead, thinking about digital worlds as emancipatory spaces is best understood as a metaphor for thinking about socioeconomic change and the pursuit of global justice, which is a latent discourse within debates about the future and technological innovation. In this respect, the Internet's main purpose is to reconstitute our beliefs about the role of the media in our societies, the consequences of which we are only beginning to see today as a result of the rise of blogging and the emerging crisis in various aspects of the media industry, such as newspaper distribution. The myth of the Zapatistas' use of digital technology is a further indication of this, as is the myth of the Wikileaks network. Each of those virtual realities was a much more modestly populated digital revolution than was thought at the height of their impact, but each has significantly shaped our understanding of how much can be achieved digitally with relatively limited means. Virtual-reality systems are like this, and we see examples of this in first attempts to create new kinds of serious VR experiences, such as *The Guardian*'s first VR journalism application, which locates the user within a prison cell using Google Cardboard, to convey what it means to put someone in solitary confinement.

Amid the claims about the value of different degrees of reality, it cannot be presumed that there is a clear instance of reality that trumps all others. Indeed, MacKinnon's (1997) definition of the real brings into question whether human existence has ever been real, since by that definition it was always mediated and interpreted. As MacKinnon states, "the primary difference then between the real and the virtually real is the interposition of some mediating and transforming agent or interface between the senses and the shared perception" (p. 4). Although MacKinnon places the condition of reality upon that which is interpreted through human senses, which is comparable to my term *sensory-reality*, this definition fails to accept its normative limitations. Thus, not all humans have the same senses or the same levels of sensory sensitivity—people have different degrees of smell, sight, hearing, and touch. There is no singular reality that all people share or which can be tested against some other virtual reality. Therefore, distinguishing between the real and the non-real is nonsensical if one does so by appealing to some other, fixed, true reality. It would be more accurate to argue that human existence, having always been mediated through human senses or some other media, has always been virtual, and that cyberspace is another medium through which humans experience reality—that is, an evolution of our sense of reality.

In the context of sport, it is necessary to distinguish between the participant and the observer in response to this claim. It might also be necessary to distinguish between a spectator who travels to the event to witness it live and a remotely connected spectator who experiences the event through a broadcast or some other technological medium. According to this characterization of VR spectators' and participants' experiences may be regarded as virtual in equal measure. One might more easily accept that a digital reality is more enjoyable than viewing a television broadcast, but not preferable to actually being there. For any virtual text (book, television, radio, and so on), the virtualness of the experience is assimilated by the viewer or listener—or what we might describe as the spectator. The spectator is brought to the real context through the mediation of relevant and meaningful symbolic gestures that bear resemblance to the physical world which we expect to exist. This is why Roberts (1992) is doubtful that representations of sport can ever do justice to the embodied experience of sport participation. However, simulation technologies may hold greater prospects to achieve a more truthful depiction of embodied action. Consider the recently launched platform Virtually Live, which uses 3D mapping and GPS tracking to translate a sports event in real time into a computer-generated moving image. The viewer can watch the sports event take place through virtual-reality goggles and see the actual movement of cars, but their view is of a computer-graphic rendering rather than of the live camera feed. In this case, with a lag of only 6 seconds between the event and the virtual experience, the simulation may be able to add additional layers of information into what's happening (additional, that is, to what we see when relying simply on our senses).

Conversely, one can speak about virtual experiences for a participant. If one is within a virtual reality and the simulation is effective, then its virtual status becomes irrelevant; it is simply another form of reality. Conceivably, there might be some residual awareness that leads the participant to realize that the experience is a simulation. For example, if one is playing a virtual-reality combat game, then even if being under attack by some opposing military enemy would not create the kind of psychological trauma that a real attempt on one's life would create, because the player is aware that he or she is in a game. However, deliberating on this point becomes uninteresting, as it becomes a matter of empirical fact as to whether the technology is able to deliver such richness of experience. Theoretically, if the simulation is good, such residual awareness should not be present and would be more like a form of psychological delusion where we are unable to tell the difference. Indeed, one would expect that in a true simulation—as was depicted in *The Matrix*—one would not be able to recognize the difference between the simulation and something that is external to it. Whether such simulations could be replicated through computer technology is, however, a matter also for technology to demonstrate. Yet we need not turn to technology to instill such feelings; we have them already in varying degrees of dream states. This alone seems sufficient to conclude that it is not impossible to create imaginary worlds that can seem real enough and which have real effects on one's personality. Consequently, from both perspectives it seems likely that creating convincing digital environments that are indistinguishable from other sensory-realities is possible.

Two very different ideas about virtual reality were advanced in the preceding two paragraphs. The former (strong VR) recognizes that virtual reality describes the entire lived experience of humanity as a virtual experience and argues that there is nothing revolutionary (and thus alarming) about new digital realities. The latter argues that, even if one does not accept the virtuality of "real" life, it is also possible to claim that digitally mediated virtual realities can be sufficiently real (weak VR), and that these too can be desirable experiences to pursue. What seems unequivocal is that there is a second wave of convergence taking place within the digital sports world.

To understand more about this shift, it is useful to revisit the ideas surrounding the first wave of convergence, which are best explained through Brian Stoddart's ideas about the "information superhighway" in sport. Stoddart (1997) imagined how various information structures and services, including home computing, television, radio, telephone, and email, would merge. That process would follow from the growing desire for more media consumption and the improved efficiencies that would be found in delivering personalized digital systems across networks. Twenty years later, many elements of that process have come to fruition, and sport has led many of them. Today's transmedia experience entails the capacity to share content from one platform to another, 24 hours a day, seven days a week, in physically stationary places or on the move. It involves the blurring of realities from game spaces such as sports to the game spaces of computers. Stoddart's convergence never really focused on how the human experience would change as a result of this technical transformation—how the user would be affected. Yet we must consider the value of new technologies from a human-centered perspective if we are to derive an understanding of their importance.

To help us understand the effect of recent digital change, we need new metaphors to capture the Web 2.0 era and its impending consequences for human interaction. We no longer surf the Web, browse the Internet, or download data. Everything is streamed in real time, located in the cloud, embedded across platforms, and persistently in transit from one space to the next. If the Web 2.0 architecture of embedded content were to collapse today, most of the content on the Web would disappear, but this diminishes as we move increasingly into an app-based Web environment.

The idea of second-wave convergence advances a sociotechnical interpretation of media change that recognizes the need for societies to go through a cultural convergence, as much as a technological convergence, in order to realize the benefits of digital change. Sport is a strong indicator of how such processes are taking place. For instance, one may speak of a convergence in the experiences of athlete and spectators, as I will elaborate in part II. A key influencer of this trajectory is the desire of media organizations and sport producers to bring spectators closer to the athlete's experience, to allow them a greater understanding of what it takes to make a great sporting performance possible. To assist in this pursuit, sports arenas are beginning to experiment with technology. For example, the billion-dollar design of a stadium for the NFL's Atlanta Falcons includes in-seat vibration technology—called "impact seating"—activated by collisions between players on the field. This technology echoes developments in other entertainment pursuits, such as the creation of the D-Box cinema experience, which allows the viewer to sit within a simulator that will move in response to what is taking place on the screen. The future of spectator experiences may thus be seen as a dynamic whole-body simulation, not just a seated experience.

This chapter has considered the idea that there is a reality against which virtual reality can be judged. It claims that there is nothing inherently valuable about either version of reality and, by implication, that nothing about a growing, digitally enabled virtual reality should concern us greatly. Sport in virtual reality might be devoid of the human body or even human contact as it is conventionally understood, though our present understandings of contact and communication are not adequate. Offline worlds do not have value that is superior to the other possible ways of being engaged in sport. Moreover, they do not lack the characteristics of sport that give it value. The human being is not absent in virtual sport. People still come together in virtual sport. Real communication still takes place in virtual sport. Any claim that virtual sport is reducing human existence to a functional, rationalistic way of living also misunderstands its possibilities.

In virtual sports, teams may not travel to a common physical location and compete on a physical track or field. The digital interface of Sport 2.0 may not be as inclusive as the physical worlds in which people gather. However, any loss in richness may diminish as people adjust to their new circumstance and as new technological possibilities are realized. Furthermore, when placed against the financial costs of bringing people together in physical space across the world—not to mention the environmental cost of getting there and the real difficulties of making the activity truly inclusive (not everyone can actually travel)—there are good reasons to aspire toward digitally delivered sports events. Historically, sports events have been imagined as experiences that must take place in physical locations. Some years ago, social events were also conceived in this way, but with more and more virtual chat locations, this mode of existence is now only one possible means of interacting with others. In short, a lack of interest in making sports virtual reflects more a nostalgia for a particular way of seeing our bodies than anything that is inherently essential to what sports should be. Such nostalgia serves to keep sports in a particular format, which is wedded to our sense of history, but it is a view that fails to take into account how we are evolving more broadly as a species. Our bodies are quickly becoming conduits of data and experiences, but the precise mode of that interaction need not be fixed.

Early debates about the Internet focused on whether it would herald the possibility of living a life of simulation, which no longer required us to enter the physical world again. Questions focused on whether people could map their understanding of the physical world onto cyberspace, replicating the same sense of what, for example, it means to be part of a community or our identity. Even today, these same questions persist about how to make sense of virtual realities, as either unreal-world environments or an integral part of our world. These questions are not asked solely of sport or forms of digital culture. Rather, one might make similar claims about other cultural or leisure-based activities, such as theater, literature, or music. Indeed, memory and dreams may also be forms of unreality that have a close resemblance to the kind of unreality present in either of the two practices under discussion. However, I have endeavored to show how each occupies a space that resonates with Michael Heim's (1993) "metaphysics of virtual reality" to understand more about how different forms of reality have been treated historically and, as a result, how one should regard them in the future. After all, it is not yet apparent that the nature of digital virtual worlds is significantly different from, say, the virtual world of the simulated sports arena or any other form of unreal space. Indeed, Manuel Castells (1996) notes that the term *virtual reality* is misleading since "there has always been a separation of reality and symbolic representation because we always interpret everything we encounter through some system of meaning." Similarly, N. Katherine Hayles argues that

For information to exist, it must *always* be instantiated in a medium. Whether that medium is the page from the *Bell Laboratories Journal* on which Shannon's equations are printed, the computer generated topological maps used by the Human Genome Project, or the cathode ray tube on which virtual worlds are imaged. (1999, p. 13)

Both Castells' and Hayles' assumptions about unrealness also lead to challenging the further claim that our experiences of the virtual world is less important or valuable than the face-to-face or embodied reality of everyday life and that this has something to do with their unrealness. In short, there is presumed to be some reality "out there" to make the mediation possible. On this view, the mediated reality is our objective and tangible context against which all other virtualities are juxtaposed. This conclusion challenges the idea that this "reality" is also virtual and would, instead, accept that there is a true reality. Such a preoccupation with virtual reality's inferiority is a foundational concern of this book, which asks what kind of importance sport has outside of its internal competition structure. Similarly, one might consider whether blogging has the capacity to transform society. In each case, the goal of the inquiry is not merely to unravel the philosophical meaning of concepts, but to understand how ideas are brought into existence through policies that place value in some things rather than others. In the case of blogging, the question does not simply ask whether words written in cyberspace can be influential outside of it. Rather, it inquires into the manner in which cyberspatial artifacts find their way into broader socio-political discourses. This is the crucial and limiting factor in determining their significance.

Despite twenty years of research arguing on behalf of the importance of the Internet as a political medium, it is still unclear how one should situate something like Twitter feeds within the media nexus, in part because it is difficult to judge which forms of mediation will last. Back in 2008, just over a year after the newsweekly *Time* announced "You" as its Person of the Year on the basis of the volume of user-generated content online, debate ensued about the influence of YouTube in Barack Obama's presidential campaign. The following year in 2009, Twitter became the focus of debate about the world's perception of the Iran elections. One reason why it is difficult to resolve these matters is that the debate about the value of social media often focuses on the platform, rather than on the professional practice that surrounds it. Similar questions are often asked about sport. In 2009, the International Olympic Committee was awarded Observer status at the United Nations, and for some years it has been preoccupied with promoting sport as a mechanism of peace promotion, but does it really bring about such changes in global political actions?

Each of these contexts—digital and sport—requires careful assessment to better understand how each constitutes our sense of reality and the social conditions that define it. From the monetization of Second Life to the widespread development of gambling communities for virtual sports fans, life online shapes and is shaped by life offline in many ways, establishing nuances in our use of the term *reality*. How these practices inhabit our lives and the various roles they play in constituting our worlds are crucial to our deciding whether or not to ignore the emergence of new practice communities in digital space.

The more sobering perspectives on the future of virtual reality that I have advanced ought not dampen our enthusiasm for its pursuit. Within sports, there is nostalgia for a pre-technological era, which is characterized by a mistrust of such innovations as virtual realities. Yet the more creative imaginations that depict virtual realities as worlds where humans interact telepathically or where smart textiles allow a person to feel as though he or she is moving in a computer-simulated world, influence how we imagine future worlds. However, these ideas should focus our attention on the world changing potential of technology and the importance of technology becoming mundane, before it can become radical.

II E-Sports in Three Dimensions

Taken together, sports and creative computer-based experiences occupy a significant part of our social world. Yet their common conceptual grounds are much neglected in studies of media and of sports. The previous chapters explained why there is value in theorizing the synergies between these areas of human experience. I now wish to argue that it makes less and less sense to talk about sports and digital cultures as distinct entities, but also to advance the idea that the recent rise of e-sports is indicative of this common future, which becomes apparent when one examines the totality of digital technologies that make sports possible. Increasingly, sports are produced through the utilization of digital technology, and digital environments increasingly rely on the logic and the ethos that are characteristic of sport.

Certainly, sports and digital worlds are experienced in remarkably distinct ways and the richness and complexity of these environments does not easily permit their conflation. Indeed, some fans or practitioners of sports will find little or no value in engaging with sport via digital interfaces. Whether the bodies are powerful or graceful, solitary or clashing, such people would claim that sport's most salient characteristic is its corporeality, and that digital technology has no place in this arena. Similarly, digital natives may have no interest in getting physically more involved with their desk-based or mobile-based pastimes, or their being gamified. However, these communities of resistance are being compelled to adjust as more of sport is delivered digitally, and the change is coming from both sports and digital innovation.

A good example of this change is how the experience of physical activity is being made central to digital experiences as a result of wearable technologies, which are changing the means by which people engage with and create digital content. Indeed, new mobile interfaces reveal how much can change still in how we design and use digital equipment. For instance, one design feature of Google Glass was its "wink" facility, which allowed the wearer to take a photograph of something just by winking one eye. Until Google Glass, taking a photograph had always involved using some form of hand-based interface and the art of photography had developed around this feature of all cameras. Google Glass, for all its shortcomings, challenged this idea and demonstrated that our assumptions about what photography entails could be different.

The integration of physical movement into our experience of computing can be traced to the early years of digital gaming. Players began to experience physical reactions in their bodies as they moved a joystick in correspondence with what was happening on the screen. Tilting a joystick, even though this had no impact on what took place on the screen, was indicative of the imminent haptic shift within digital culture. In this respect, one can observe how sport and digital technology aspire to similar goals, namely the creation and experience of alternate realities by experimenting with forms of simulation. Indeed, the simulated dimensions of sport run even deeper as, on one interpretation, they are wholly organized to permit a more civilized playing out of what takes place elsewhere in the form of war and violent conflict, which serve to establish hierarchies and power relations.

Today, sports with civilized rules entail the simulation of various tests of human capacity, as a preferable alternative to actual violence. Similarly, digital game players—or e-sports athletes—do not typically wish to train for years before being able to use a flight simulation computer game. Instead, a player's aspiration is to experience a simplified version of such skill development and acquisition. The 21st-century game-playing world is being defined by simulations, which are becoming indistinguishable from the simulated world they depict. Increasingly, pilots are trained within simulators, the sophistication of which is becoming affordable to consumers. In 2008, a renowned e-sports player was appointed to Nissan's auto racing team as that company's first "gamer to racer." Furthermore, platforms such as iRacing (see iracing.com) are working with official motor-racing teams and competition organizers to create computer-generated models of cars within Games that are accurately constructed in terms of design, physics, and geography.

Yet there is still resistance to such convergence. Typically, people do not regard computer games and sports as occupying the same kind of onto-logical space at all. Another kind of resistance toward treating sport and digital worlds as similar comes from spectators who reject the value of remote viewing, either through television or the Internet, arguing that the experience is richest when one is live. Yet the digital viewing experience is becoming even more visceral and more engaging than the live experience as 3D high-definition television develops alongside virtual-reality simula-tors. One future prospect for the convergence of sports and digital worlds is the development of seamless simulation machines, which may digitize participation in sports so thoroughly that the difference between digital and physical participation becomes imperceptible.

Beyond a keyboard or a hand-held interface, the exotic allure of virtual-reality technologies involves the use of three-dimensional acoustics and stereoscopic optical displays, head tracking devices, data gloves, body suits, and telepresent robots to add sensory feedback, ambience, and movement control options. Many of these advanced technologies, as they have been realized, differ from the ways they were imagined in the 1980s. Today *virtual reality* means something very different from what was expected even twenty years ago, and it will be useful to reconsider today's VR in a post-digital world (that is, one in which there is more concern with staying human than with becoming more digital). Indeed, the turn away from vir-tual reality to augmented reality and even "mixed reality" is indicative of the shift away from life in a virtual world and toward a physical world that is enriched by digital artifacts. On this version of the future of VR, it is far better to add layers of content to the physical world than to do away with it completely. We see glimpses of such possibilities in recent prototypes, such as the RideOn augmented reality goggles for snowboarders, which overlay digital gaming content onto the physical world, allowing novel gaming experiences to take place. For example, snowboarders may be able to visit the slopes at a previous Olympic site and bring up the slalom run within digital head displays, allowing them to test their skills against those of for-mer Olympic athletes. When applied to games and sport, VR technology often involves transforming users into virtual players, or the wearer acts as a telepresent agent for robotic players. When life in virtual worlds comes to dominate our daily interactions, playing sports in the physical world may have no or little appeal once new communities of cyberathletes have taken ownership of sport. In this scenario, the future of sport is defined less by resistance to the virtual and more by a lack of interest in non-virtual artifacts or spaces. Whereas once life imitated art, it seems now that life offline is beginning to imitate life online.

The chapters in this part of the book address such prospects from the perspectives of athletes, spectators, and officials—those who are engaged in the production of the theater of elite sport. Yet their trajectories are not wholly about technological change; they are also about understanding what simulation requires of us culturally. For instance, it is not enough to simulate an arena to satisfy the interests of committed live spectators; it is also necessary to allow them to enjoy the social experience of sitting in a sports arena among other spectators. When we are at a sports venue, watching the action, it matters that we can interact with others and that we can interact with all elements that surround us. In this respect, creating good simulations requires understanding the reality of a spectator's sensory world, of which watching and hearing the competition is only one component. These complex cultural aspects of the sports experience cannot be overlooked when proposing the idea of simulation as a closed system of computer-generated experiences. In short, one cannot simplify sports too much when aspiring to simulate them through digital technology.

Understanding the environmental noise that operates around a sport involves situating sports experiences within their broader social context. It requires understanding what participation in sports involves and how it is valued, which are often intimately informed by our experiences of work and play, as well as to a range of other social categories, including gender, health, age, and ethnicity. Our perspective on which of these factors are most relevant to emulate will differ, depending on our priorities. For those who experience sport as a commercial endeavor, it is an increasingly essential part of the intellectual property rights that surrounded many other creative industries. For those who are focused on sport's community value, its primary good involves reinforcing bonds within social networks or the family. Understood in such terms, sports competitions and the results that follow from them are of secondary importance to a fan's social experience, even if the play and the results are the focal points of that experience. Indeed, such a view explains some of the anxiety about certain sports behaviors, such as cheating or the violence that they involve, since these manifestations of action seem to compromise the aspiration for sport to reinforce positive socialization.

These interpretations of sporting and gaming cultures add to the complexity we are presented with when attempting to understand the past, the present, and the future of Sport 2.0. We live in times of new collaboration across digital and sporting cultures. The term *Sport 2.0* is symbolic of how these areas of human experience may come together while also denoting the emerging professional e-sports community, constituted by vast communities of sports gamers around the world. Sports gamers are provoking those who are in charge of traditional sports to consider whether they should embrace a new generation of athletes who are digitally constituted and committed to physical activity. Beginning with changes in the experience of elite athletes and then proceeding to consider developments in mobile health technology used by amateur athletes, the chapters in this part elaborate on how the idea of Sport 2.0 reveals a vast digital architecture that underpins 21st-century sport. The ubiquity of digital technologies is the reason for this change, but the desire to inhabit simulated worlds is as much a part of sport as it is a part of digital environments. Taken together, the ideas presented in this part of the book reveal how the present definition of e-sport is still embryonic in its development and that digital gaming is becoming a bigger part of the elite sports environment in many ways.

3   The Digital Ecology of Elite Sports

This chapter considers how digital technology has altered the world of elite athletic performance and what this means for the future of sports. More broadly, it explores how digital technology has become a pervasive—and legal—form of performance enhancement, along with having become a ubiquitous presence in an athlete's life. In so doing, it investigates how digital technologies have altered how athletes train and how they impact upon the fairness of sports, while also considering how this has influenced the work of those officials who oversee the smooth running of sports.

Today, many sports training programs use various forms of digital data capture to model complex sport movements. Data generated by new digital sensors helps to reveal the consequences of tactical decisions, or to understand the physical limitations of athletes, so as to provide insights that will make it practicable to enhance athletes' performance. Whereas the different forms of data once were discrete, today digital technologies are increasingly holistic and integrated, involving monitoring all aspects of performance and permitting a more comprehensive aggregation of information that can inform decision making about an athlete's performance development. It is in this respect that 21st-century digitized sport is part of a wider interest in big data and the desire to exploit it for economic gain.

Despite the value of these achievements to sports performance analysis, trends toward digital ubiquity may be criticized for their potentially dehumanizing impact on sport's physical culture. For example, "third eye" refereeing has been discussed by academics and sports professionals as having the potential to undermine the role of human judgment in refereeing (Collins 2010). Alternatively, subjecting talent identification to digitally automated filtering systems professed to have the capacity to reveal talent potential as a statistical probability may compromise the role of complex judgments made my talent scouts, diminishing those aspects of sport that rely on an expert's intuition about what makes an athlete likely to succeed.

Yet there has been considerable progress toward intelligent systems—and a loss of confidence in human expertise—in the past two decades. Moreover, these developments fit within sport's aspiration to achieve greater precision in the analysis of performance and the development of sport science as an industry, the roots of which may be tied to a wider scientization within sport (Guttman 1978; Hoberman 1992). The increased use of digital technology—especially as an analytical tool—is an extension of sport's teleology, where less and less about an athlete's performance is left to chance or human error. Indeed, in this respect, it is presumed that intelligent, algorithm-based refereeing systems ensure that fewer errors are made; however, their merit relies on the intelligence of the model underpinning their design, and that may be where the flaws reside.

For some, the prospect of replacing human judgment with automated intelligent systems' describe a dystopian view of humanity's future that entails a failure to recognize the complexity of human embodied intelligence, which cannot or should not be reducible to an algorithm. On this view, such knowledge is important to retain as a form of uniquely human knowledge as it is understood only by experts who have undertaken the necessary encoding into their corporeal memory that such a journey allows. These presumptions about what constitutes sporting knowledge—and knowledge more generally—explain why coaches and managers tend to be former professional athletes: their knowledge is predicated on the embodied insight they derived as practitioners. Moreover, the mythology of sport relies on the belief that the athlete is, in some way, a person with rare gifts whose achievements are attributable to unique and otherwise inaccessible insight into how to perform within his or her body.

Resistance to reducing human judgment to an algorithm is not the only kind of technological resistance in sport. Indeed, the sports community is often noticeably opposed to technological change more broadly, partly out of a concern that new technology will corrupt some aspect of the internal value of a sport. Trabal (2008) takes this idea further, highlighting the complex journey taken by any new sports innovation, where it first meets various structures of resistance before any possible acceptance. Trabal argues that the inherent value of the innovation making a sport better, or a performance more efficient, is not the primary means by which one can explain its success of failure. Rather, there are different, dynamic factors that can play a part in the success of any given sport technology, which includes the readiness of its supporting infrastructure, or the political support it receives within the practice community. Nevertheless, the concern that technology may jeopardize something that has been created, codified, and institution-alized is a significant factor in explaining technological resistance—and digital technology is a case in point.

A good non-digital example of this was the challenge posed by the South African Paralympic champion Oscar Pistorius, whose prosthetic lower limbs were seen as a challenge to the rules overseen by the International Associa-tion of Athletics Federations (IAAF). Pistorius' aspiration to compete along-side so-called able-bodied athletes was challenged by the IAAF for similar reasons—that his technological aids undermined the integrity of the com-petition. In technical terms, the argument hinged on whether or not the prosthetic limbs provided a kind of momentum that meant the way Pisto-rius ran was biomechanically different from how a person with biological limbs ran. If so, then the sense of the event was lost, and it was not deemed acceptable that one athlete had an unfair advantage over another in this way.

This example of resistance to technological change invites us to consider what it is about human performance that matters. Does the reduction of all aspects of sport to some scientific principle diminish the value of the performance, or can it promote the celebration of other values? There is reason to urge caution in the critical view of technology, not least because the value one attributes to present-day elite sports performances is at least partly predicated on the same scientific processes that critics question. In other words, if one acknowledges the value of athletes' refining their practice and getting better at what they do through technological means, then it is difficult to oppose technology, as, over time, it becomes an even more crucial part of making such achievements possible. Indeed, following in the footsteps of Pistorius, the German long-jump Paralympian Markus Rehm has emerged as the next generation of prosthetically enabled athletes whose capacities may allow him to win medals in the Olympic and Para-lympic Games.

Digital technology reveals aspects of an athlete's insight that might oth-erwise be attributed to intuition or talent. By implication, although digital technologies may diminish the mythology that surrounds the performer, they also demystify what it requires to perform at the level of an elite athlete. Yet this is also why technology can be so controversial. After all, there is a sense in which technology can reduce magic to mere mechanics, and it isn't clear whether or not such explanations enrich our lives. One crucial aspect of the debate concerns the value attributed to physical-world experiences. Thus, in contrast to how spectators experience the impact of digital technology, the direction of travel for athletes is found in its potential to make the conventional sports arena redundant as sports become enterprises that are practiced increasingly within virtual worlds rather than physical spaces. Consider the rise of the popular Les Mills Immersive Studio, which surrounds a spinning class with a 360° screen displaying computer-generated imagery of an imaginary cycling route.

In the future, highly sophisticated ergometers may replicate the competition environments entirely, and 360°, holographic, or 3D digital broadcasting may construct the arena for the viewer making the stadium redundant. Additionally, such technologies will create the conditions of an athlete's experience as a performer, isolating the act of physical movement as the main criterion of what sports seek to measure when placing athletes alongside other athletes in competition. Thus, one might envisage cyclists competing within simulated worlds, sitting on real bicycles—not too different from exercise bicycles of today perhaps—but cycling within an entirely digitally constructed terrain, carefully designed to optimize the competition, offering the most precise comparisons between athletes from one tournament to the next. The input of the cyclists' efforts would be digitized for the athlete and the viewer, communicating their performance in real time. As for present-day technology, the athlete's performance data could be visualized alongside other data belonging to other athletes in the race. Examples of this are beginning to emerge, through such companies as "Activetainment," which combines a cycling simulator and virtual-reality goggles to provide the athlete with a near-world experience of cycling in remarkable terrain.

Some people may recoil at the idea of moving sports out of physical arenas and into virtual worlds, or criticize this scenario as an excessively reductive view of what happens within the competitive sports environment. From this perspective, it is not just the conditions of the playing field that make sports meaningful experiences, but the entire range of interactions they elicit, which more broadly describe them as a cultural configuration and not simply a competition. This holistic view of what happens in a sports event encompasses the physical proportions of the space, an athlete's temporal journey through it, the sounds and smells of the environment, and the sense of proximity to other people, or what may broadly be described as the "atmosphere" of a place (Chen, Lin, and Chiu 2013).

Reconciling these two perspectives on sport's digital future relies on the ability of simulated worlds to deliver compelling realistic simulations, without involving any significant sense of loss in comparison with what we get from physical-world experience. Yet it also requires a sophisticated understanding of what such loss entails and whether it matters, and there is evidence to suggest that we do not quite have such an understanding with regard to our comprehension of life online. For instance, the concept of loss within social-media environments specifically and virtual worlds generally is contested. In some psychological studies, participants in virtual worlds were found to feel no sense of obligation if made aware of potential risks others might face. My claim is not that a similar kind of relationship to others ensues in virtual worlds, but that there is sufficient social communication to support the argument that people find virtual experiences enriching. Even if one concludes that interactions in the physical world are preferable to interactions in the virtual world (a view not supported by the literature), one's inability to create physical worlds with such dynamic and physically present dimensions is suggestive of the additional value one may derive from life online. Were technology to achieve a state in which there is no sense of loss perceived about presence within physical space as opposed to offline worlds, the contrary view might apply—that the virtual worlds become even richer than the physical-world experiences because they can include additional sensory content that exceeds the possibilities of the physical world; such is the goal of augmented-reality technology.

A further rationale for rejecting digitally enriched future sports is that sport is already sufficiently exciting so as to not require augmentation. Yet there may be circumstances in which this view is jeopardized and where digital technology can provide additional ways of differentiating and heightening experiences. For example, consider sports in which distinguishing finalists is difficult because the athletes are all very similar in their capacities. In such sports, technology may provide new ways of discerning difference in a similar way to how photo finishes enabled such differentiation.[1] There may already be a need to measure foot races in thousands of seconds, rather than hundreds. Although Baldwin (2012) argues that it wouldn't be fair to distinguish first and second place on this basis,[2] the challenge for sports is that first and final places may be indistinguishable in the future, raising questions about how else to adjust sports to allow a clear winner to be identified. In this scenario, digital technology may be the only means by which we enhance our perceptive skills to make sports interesting again, to show us who has won or to judge victory differently. Alternatively, presumptions about the sufficiently rich and immersive nature of present-day sports may be undermined as the spectator experience changes. In the past, a spectator may have simply sat in a seat in an arena and watched the sport. Today, many spectators hold mobile phones and cameras as they watch, and interact with content beyond what is happening on the playing field. These circumstances remind us that the nature of immersion can change dramatically and is technologically dependent.

There are various aspects of the elite athlete's environment that demand closer inspection, and in subsequent sections within this chapter I aim to tease out these nuances. First, I will focus on how digital technology affects an athlete's embodied experience of training and competition, considering how the skills required by athletes change as a result of these developments. Second, I will consider how officiating and surveillance of the sports field are affected by digital technology, notably in the way that equipment is used to enforce the rules and eliminate the human error that is implied by relying solely on an official. For each sport, the use of digital technology operates in varying capacities, though in all cases it promises to have a far greater influence on the interface through which an athlete competes. Finally, I will address the increased role of social media in an athlete's career and how this presents new challenges for their professional conduct.

Digitally Trained Bodies

Digital technology finds its way into the performances of elite athletes most often through training methods and the capture of performance data. In the past thirty years, digital technology (for example, timing technology, heart-rate monitors, exercise ergometers, and virtual-reality simulators) has dramatically transformed how athletes have trained. Its impact continues to grow as greater methods of modeling sports fields are developed. Though visual data capture has been used in sport for many years in analog times

(McGinnis 2000), one of the earliest examples of digitalization within sport is the recording of performances. Precise digital measurement has become more influential as athletes have become more competitive and as it has become more difficult to distinguish first from second place. During competition, the rules preclude certain kinds of digital interventions. For example, an elite athlete is not allowed to use an iPod while running the New York Marathon, because a musical rhythm may help to set a pace. This prohibition is specified in USATF Rule 144:

The following shall be considered assistance. ... The visible possession or use by athletes of video, audio, or communications devices in the competition area. The Games Committee for an LDR event may allow the use of portable listening devices not capable of receiving communication; however, those competing in championships for awards, medals, or prize money may not use such device.

However, outside of competition, many athletes have pursued performance enhancement through developing sophisticated digital enhancements to training methods. Digital technology may even be used in pre-training situations to study the activity of potential elite athletes (Ives et al. 2002). For instance, FusionSport (2012) claims that "over 90% of the Universities in Australia involved in the Australian Institutes of Sport's eTID program use Fusion Sport technology for their testing"—that is, they use it to identify new talent from large groups. FusionSport's wireless-based timing technology provides a level of precision that is characteristic of the requirements of elite sports today, which reinforces the need for digital solutions to allow athletes the possibility of performance improvement.

Within the broad spectrum of athletic performance enhancers, digital technology falls into the category that the technological determinist Jacques Ellul (1964) described as "La Technique": the collection of forms of knowledge that underpin any particular technological artifact. Ellul employed a definition of technology that went far beyond artifacts. This technological category also encompasses the knowledge gained through scientific endeavor, including databases of information that may be used to interpret performance achievements. For Ellul, technology also adhered to a rationalist logic that we may extend to interpreting the history of modern sport as a scientific project in its pursuit of more refined, more precise, and more extraordinary performances. In the early 21st century, modern sport began a journey which tied it to technology as the route through which progress would be achieved, and digital technologies have become intimately connected to these wider configurations. In this respect, digital technology should not be seen as isolated from other forms of technology. Rather, its utility is derived from other systems that produce meaningful digital data. For example, the "ski robot" created at the Jozef Stefan Institute in Slovenia may yield insights into robot design that will have applications outside of skiing. Indeed, the goals of such research may not be tied to elite sports at all, but may more broadly aim to "imitate human behavior and thus make a robot a useful assistant in everyday life" (Lahajnar, Kos, and Nemec 2009, p. 567). Yet transfers from such designs and research programs into the world of sport have happened in sports for decades.

Digital training is most often utilized as a form of feedback for athletes (Liebermann et al. 2002; Liebermann and Franks 2004). Some of its uses entail computer-aided analysis of player/team movements and body positions (Perl and Memmert 2011; Macutkiewicz and Sunderland 2011), computer-based models of movement interactions between athletes and equipment, or simply tracking performance activity during competition. For instance, ProZone allows the tracking of soccer players, which shows a positive correlation between ball possession and sprinting. Thus, the more a player sprints in the course of a game, the more likely he or she is to obtain possession of the ball, which can translate into greater success within the match (ProZone 2014). Each of these applications may utilize a range of delivery platforms, from mobile devices using GPS to virtual-reality simulators.

One prominent early example of an advanced digital training technology was used by the USA Olympic bobsled team in preparation for the 1994 Winter Games (Huffman and Hubbard 1996). A later virtual-reality simulator replicated the Nagano 1998 Olympic Winter Games sled run. Such simulations can be particularly valuable in situations where competitors have only a very limited number of opportunities to practice on the physical terrain before competing. Indeed, the value of using such simulations before teams arrive at the Olympics was made explicit at the 2010 Vancouver Winter Games when a Georgian luge athlete, Nodar Kumaritashvili, crashed and died during a practice run. Though the inquiry into did not find fault with the design of the course, that unusual accident raises the question as to whether competitors are given adequate opportunity to learn a new route before competition. With bobsled and luge in particular, there is value in learning the twists and turns of a course, and the USA team's simulator enabled athletes to do so. They could use the simulator to build their knowledge of the bobsled course before traveling to Japan and seeing the completed track. A more obvious solution would be to permit further practice runs before competition, but a time-critical schedule may preclude sufficient opportunities. A similar approach to helping athletes become familiar with a competition environment was explored by Sorrentino et al. (2005) with a group of Canadian Olympic speed skaters, and the results of their experience indicated that virtual environments promote athletes' visualization in preparation for competition.

One exciting aspect of simulator technology today is that consumer products have become incredibly similar to those simulators used with elite athletes. For example, the VROX entertainment simulator allows users to experience a bobsled run just for fun. The overlap between these entertainment experiences and what might be useful as a form of training for professional athletes is getting stronger. The growth of simulators and goggle technology such as Oculus Rift is beginning to show a new range of ways to simulate sport experiences, and not just for spectators.

Coaches can also monitor athletes during competition by means of wearable technology such as SporTrack, Trak Performance, and inMotio, which enable real-time tracking of multiple performers that can provide sophisticated data about how athletes are performing individually or as a team (Barkett 2009). Film capture using external devices such as SportsCode, DartFish Team Pro, or SportVU provide further means of data-driven performance assessment. Monitoring athlete data often operates in tandem with the analysis of how athletes interact with the equipment they use. For instance, Akins (1994) discusses how the United States Golf Association utilized technology to assess the properties of new club designs to ensure that players were not given too much of a performance boost by the latest innovations. Indeed, in the past twenty years golf's encounters with new club and ball technology have provoked considerable debate over whether the game is being transformed in a way that undermines its integrity. Moreover, Akins' forecast that "tomorrow's golfer will be able to buy a set of clubs that has been designed according to his or her needs and abilities" (p. 41) has been a reality for quite some time, thanks to the use of computer aided design. More recent innovations allow golfers to track their ball using GPS technology. For instance, GolfLogix provides players with insights into their performance by translating game play into visually detailed maps of the course, providing information about the player's ball's flight path, distance achieved, and progress from one round to the next.

The utilization of digital technology for training purposes coheres with its main premise, which is to prepare the athlete for competition by closely modeling those conditions. Thus, the more seamless the training experience, the more likely it is that the rehearsal will prepare the athlete for competition. Of course, the challenge for technology is to achieve a precise replication of the competition environment. As Liebermann and Franks (2004) noted, not all simulators do this equally well; some cause motion sickness that compromises the entire purpose. Sometimes, when attempting to simulate one reality, an entirely new reality is created instead.

iReferee: Could Artificial Intelligence Do a Better Job?

Another aspect of the athlete's experiences in competitive sport has to do with how their performances are regulated by digital officiating systems. Often, debates about the need for such innovation in professional refereeing arises from high-profile controversial results, which attract considerable media attention and which are occasioned by the failure of human officials to correctly determine what is happening within the field of play. For example, in 2010, during a World Cup soccer match between England and Germany, a decision was made whereby a goal was not given to England. The replay clearly revealed that the ball was over the line by some margin, but the referee and linesmen failed to see that at the time. The incident reignited a long-standing debate within the world of soccer over the need for goal line technology (GLT), which uses high speed cameras and timing technology to record the ball's location. The 2010 World Cup incident was one of a series of crucial errors spanning more than a decade, all of which could have been prevented had the use of goal line technology been employed. Perhaps because of England's status as a powerful football nation coupled with it being a high stake game, on July 5, 2012 the International Football Association Board approved the use of GLT. At the 2014 World Cup Finals in Brazil, goal line technology led to the first result being decided on the basis of the GoalControl system (Chowdbury 2014).

In some sports, what happens on the playing field has changed dramatically in recent years, making it harder for officials to do their job without technological assistance. One of the best examples of this comes from men's professional tennis, in which the average speed of play has increased so much that being able to determine whether a ball is played in or out is harder today than it was thirty years ago. In each of the Grand Slams, the first serve speed in the men's game varies, but for the French Grand Slam it was approximately 160 kilometers per hour in 2000 and nearly 190 in 2008. The increase is comparable for other Grand Slams (Cross and Pollard 2009) and is one reason why tennis has pioneered digital officiating systems such as Hawk-Eye (Duncan, Thorpe, and Fitzpatrick 1996, p. 22). However, its introduction at major tennis tournaments (in 2006), cricket matches (in 2009), and British Premier League soccer (in 2013) was not without controversy, not least because of the large costs associated with installing the equipment.

In many sports, the wider objections to reliance on digital officials are twofold. First, there are concerns that appealing to a machine may automate a part of the competition environment that is best left to human judgment (Collins 2010). On this view, the referee is not just an enforcer of rules, but a crucial arbitrator of the ambiance of the match—perhaps more like the conductor of an orchestra than someone who applies the rule book. A human referee will know how severely to enforce a rule and, at times, may even choose to allow a match to continue without awarding a penalty, even if the rules require that a sanction is warranted. The second reason behind objections to the use of digital technology to assist officials has to do with whether their use will interrupt the flow of the game and prevent the demonstration of abilities that people feel are central to the sport's value. This view takes into account that what happens within a sports match is a fine balance of circumstances that elicit an optimal range of opportunities to demonstrate skill and create a degree of tension that can be easily spoiled if small changes are made. For example, if camera technology were used to support every decision a referee made in soccer, the game would be brought to a standstill far too often.

The growth of the sports industries and the amount of money they generate create a great deal of pressure on sports federations to ensure that the right decision is made about any technological change. For that reason, it is hard to envisage a future in which resistance to digital officials finds much support, especially if its use is limited to crucial decisions within the game. While determining what constitutes a pivotal decision over the course of a match is contested, sports federations have already begun to implement the technology but also limit this use. For example, in tennis, the Hawk-Eye device is used only in cases when the line call has been challenged by a player; in soccer, it is used only in goal-line disputes.

Other digital systems are also affecting the regulatory conditions of sport, some of which may require advanced innovation. A good example is referee microphones, which allow the listening audience to hear what is being said to players by the referee, as it happens. In the past, spectators—whether live or remote—weren't likely to know what players said to other players or to officials. Now everyone can hear, and this reduces the distance and isolation of the playing field to the spectators, subjecting the interactions which occur within it to the scrutiny of all onlookers. Allowing the voices of athletes and officials to be heard within the field of competition may be seen as a radical change in how people relate to their athletic heroes and may also enact a new form of disciplinary process toward players, insofar as their awareness of being broadcast shapes their behavior toward one another.

Many such technological changes to sports can also affect who are the winners and who are the losers, and this is one reason why any kind of innovation is significant for a sports federation. For instance, design changes to the javelin in the 1980s transformed the requirements of the sports so greatly that a new generation of throwers emerged. Sports are replete with examples of how technologies re-skill an activity. Yet the long term consequences of such change will not always be apparent, at the point of introduction. Edward Tenner describes this phenomenon in his 1996 book *When Technology Bites Back*, referring to J. Nadine Gelberg's 1995 study of how the development of plastic helmets in American football decreased head injuries but led to a greater perception of safety among athletes and to greater risk taking; the consequences were an increase in other kinds of serious injuries sustained in the practice of the sport. In this case, a technology that was supposed to make a sport safer simply changed the kind of injuries athletes sustained. In this respect, digital sports technologies may be situated within a broader understanding of how technology is seen as a challenge to a sport's ethos and values, without any obvious overall benefit.

Another example of such technological journeys is the design of the ball for World Cup soccer, which generates headlines ahead of each tournament, creating controversy over whether a new design modification will improve the game or make it worse. Over the years, new ball designs have been variously criticized for making goalkeeping more difficult, furthering the idea that a redesign is intended to increase the number of goals and make for a more exciting competition, presumably to make matches more entertaining. A press release by Adidas said this about a new ball design for the 2010 World Cup competition:

"JO'BULANI" also features the newly developed "Grip'n'Groove" profile which provides the best players in the world with a ball allowing an exceptionally stable flight and perfect grip under all conditions. Comprising only eight, completely new, thermally bonded 3-D panels, which for the first time are spherically moulded, the ball is perfectly round and even more accurate than ever before.

Coach Fabio Capello of England's team was reported to have said that his players were unhappy that the ball was faster and lighter than previous balls. Yet why should that have been of any concern to spectators or to players? After all, as long as all teams are using the same ball, the competition remains fair. One of the reasons for concern is that, by changing the conditions of competition, governing bodies may compromise the tacit contract which stipulates that the competition would operate according to the rules agreed to by all parties—in this case the design properties of the ball.

To understand how grave it might be to change the design of a football ahead of a new tournament, suppose that FIFA had introduced a soccer ball that was the size of a tennis ball, or one with the shape of a rugby ball. What reactions would have been appropriate and reasonable then? Fans would have claimed, rightly, that such a transformation would change the conditions of the sport in such a way as to make it a completely different test of abilities. Athletes would have claimed that they have trained for a particular type of contest and that the new design created a different one, in which they had not agreed to play. Such concern about technology is less about fairness between competitors than about the legitimacy of a governing body changing the conditions of a practice without the agreement of its members. Yet what often makes these matters even more complicated is the widespread disagreement about the impact of the change. Of course, the principle underlying the redesign of a ball is that the new model should be more consistent and closer to the perfect technological artifact for the job than the old one. In the case of World Cup soccer, any new design that can demonstrate a closer approximation to what players imagine to be the perfect ball appeals to values that all players should share, and on that basis there is a strong case for its acceptance and for this priority to trump other concerns. Regardless of the ethical debate about technological change, the transformation of equipment reminds us that sports are not static. There is no single version of any sport that persists over time, without some kind of technological change.[3]

From this example, one learns that what matters about sports technologies is that they should always perform in as predictable manner as is possible. This is why we see balls changed at regular intervals in some sports, such as tennis, in which a ball's continued use can diminish performance. Yet it is also interesting to note that we also witness athletes using this to their advantage—in tennis, for example, players may choose to serve with the same ball each time, for as long as they can. In such moments, the athlete is making a strategic judgment about which ball responds best to their skills. Even if absolute predictability cannot be ensured over the course of a competition, it is reasonable to pursue new designs to optimize predictability, in the hope that the sport eliminates the impact of irrelevant chance arising from technological design flaws. However, if a technology is introduced so late that players don't have enough time to adjust to it, or if the new technology is too radical a change, the athlete's preparation may be undermined, creating circumstances in which the technology cheats the athlete out of a previously agreed upon set of conditions.

These examples reveal how technological change creates new forms of inequalities in sport, but equality is not the sole concern of sports federations. Indeed, it is possible that new technological designs have nothing to do with fairness at all. More cynically, one may argue that publicity around sports innovations is strategically designed to generate more interest in events, and that headline stories about innovations are little more than marketing devices which serve a series of financial and political interests. Yet, aside from how new technology affects the media's narratives about sports events, there are very real consequences to changing technologies in sports, which have a bearing on what we regard to be fair. However, the challenge in understanding the implications of these changes is that their full impact may remain unknown for some time to come. Evidently, different sports treat new technologies in different ways. In some sports, athletes are given the opportunity to choose their preferred apparatus, providing it adheres to a set of specified guidelines—for example, football players can choose their shoes. In other sports—for example, World Cup soccer—all players are expected to use the technology that their governing body approves.

In most cases, what unites new sports technologies is their common pursuit of reducing the uncertainty brought to the playing field by unforeseeable environmental changes or imperfect technological design. This is why when a soccer ball unexpectedly hits a tuft of grass in the field and bounces over the goalkeeper into the goal, spectators marvel at how unlikely the occurrence was but ultimately feel a sense of injustice at its having happened. The same is true of goals by deflection, which differ from when a player unexpectedly exhibits an unusual and unpredictable performance, such as a remarkable maneuver around players leading to an exceptionally beautiful goal. When a goal is scored by deflection, the unpredictability of sport is celebrated by the beneficiaries of the deflection and criticized by those whose fortune has been prejudiced as a result. Yet such chance occurrences are broadly seen as something not to be valued, even if one may feel that the occurrences is something extraordinary. Indeed, this is why a badminton player, having won a rally with a shot where the shuttle clipped the top of the net and tumbled over to the other side, apologizes to the opponent, knowing that the victory was due more to chance than to skill. In some sports, this is even acknowledged by replaying the point.

Perhaps the best thing about a surprising new sports technology is that it offers fans some explanation for their team's failures which will permit them to enjoy a renewed sense of hope when the next matches occur. However, the choice for sports today is not to resist or embrace digital technologies, but to consider how many layers of digital surveillance may be necessary to ensure accuracy of results. On the third day of the 2012 London Games, one incident revealed just how far a technological approach to officiating may be taken. A South Korean fencer named Shin A-lam was engaged in her semi-final match, competing with the reigning champion, Britta Heidemann of Germany. With one second remaining on the clock and the match tied Heidemann needed to make one "clean touch" to secure victory; otherwise Shin would go advance to the final. Heidemann made two attempts to score the contact, without success. However, the clock indicated there still remained time for another attempt.

On her third attempt, Heidemann made a clean touch. However, more than a second had elapsed. After a 70-minute inquiry, the International Fencing Federation concluded that the decision in favor of Heidemann was fair. However, later evidence revealed that either the timing system or the referee had failed. This example goes to the heart of the debate about whether replacing—or even just supplementing—a referee's decison with digital technology is likely to improve or make sports better. In this case, in contrast with most debates about the use of "third eye" technology, the failure appears to have been attributable to a digital system. Thus, this episode occurred in a context in which technology was already providing high accuracy.

The proposal that emerges from this example is that there should be a second layer of digital surveillance to watch the watchers. Instead of using one isolated timing system, aggregating the results of two or more timing systems may be the best way of achieving an accurate result. Yet there could be other solutions. In some cases, non-technological alternatives have been proposed—for instance, human goal-line referees in football ("fifth officials") whose sole responsibility would be to monitor the goal line, rather like a line judge in tennis. This possibility also alerts us to two additional factors in deciding how best to regulate the playing field: the cost of the system and the possibility that the most effective technological solution would be to make more use of human intelligence.

In the lead up to the Tokyo 2020 Olympic Games, organizers are experimenting with the idea of using highly sensitive body sensor suits, which would convey contact in fencing to television audiences using a form of augmented reality. Multi-sensor platforms combining audio and visual elements may thus be the most effective solution to such problems. And so, in answer to the question about which is the more intelligent form of referee—a human or a computer—there seems merit in considering that a series of computers working in parallel with a referee would be the optimal solution.

The Socially Mediated Athlete

The final dimension of this chapter concerns the digital technologies that surround the communication environment in which athletes find themselves today. New channels of communication enable athletes to reach their public in different ways than were possible before and this creates new challenges. Before even considering the experiential aspect of the challenge of new and social media, it is useful to note that such changes have implications for how one makes sense of the economics of sport and the role of the larger infrastructure of sport. A good example of this is the emergence of the IOC's new "Olympic Channel," which I discuss in length later in the book.

In this section I want to examine how social media and other participatory media are changing the elite athlete's experience. For instance, there are already indications that there is a growing expectation that an athlete must be present within social media, but this can have an affect on how an athlete undertakes his or her work. Indeed, around the London 2012 Olympic Games, CEO of the London Organizing Committee for the Olympic Games (now president of the IAAF), Lord Sebastian Coe, raised concerns about athletes being distracted from their sport by social media. Yet the integration of social media into sports experiences is a growing trend. Intimations of such change are found in the emerging economics of professional e-sports, which are bypassing television and live-streaming matches to audiences directly, often through the athlete's own Twitch channel.

In view of these changes to the media economy of elite sports, federations are under pressure to re-think what takes place around their sports, to ensure they remain relevant for a generation of mobile-only media viewers, but much remains uncertain about what audience expectations of sport will be like. For instance, at some point soon might audiences expect their athletic heroes sharing their feelings *during* competition, perhaps by means of Twitter? This seems rather unlikely, but the absurdity of the proposition really depends on its detail. Of course, a cyclist isn't likely to pull out a mobile phone while in the middle of their race and share an update on Twitter. Indeed, the Tour de France forbids using a phone during the race by its riders.[4] Yet there are two ways in which one might imagine this situation differently, which make it a more credible scenario. First, wearable technologies may provide a different kind of interface for athletes to use when attempting to communicate how they are feeling during a competition. Second, data can function as communication insights, and in the future ways of expressing how we feel may be more reliant on data correlations than on direct assertions. The importance of this second feature of Sport 2.0 should not be understated, since it helps explain digital technology's potential to transform sport. The translation of sport into data leads to an expansion of the range of materials available to audiences. For instance, one tends to think of what goes on in the athlete's body and mind as unavailable to spectators, but what if we could discover it in real time by converting body data into indications of, say, emotional states? It is in this manner that sport's digital renaissance will become complete—when digital information finds ways of extending our sensory limits, providing insights into the world which we have not had before.

Again, in e-sports, this ethos of communicating while playing is built into the infrastructure of many games. Players have conversations within the platform as a competition is taking place. The appeal of this is also found in the endless pursuit of sponsors to monetize sports content further. If tweeting during competitions seems too absurd to imagine, then how about tweeting during time-outs, or during half-time breaks, which could become spaces for interaction between athletes and fans? Could sponsors contractually oblige their athletes to do a degree of in-competition interaction to heighten the experience for fans, in the same way that victors are herded to take part in certain high-priority interviews? Athletes already undertake such communications immediately before or after competition, and the pressure on them to nurture their celebrity may compel them further in this direction. Certainly, this may come at a loss in focus on the competition, but if the consequence is a more engaged audience and, by implication, a more lucrative business, then this may be a reasonable sacrifice to make. The central point here is not that sports may have to change to satisfy their need for economic growth, but that the mythical performance of athletes can be better understood through digital technology. This is appealing precisely because the heroic achievements undertaken by elite athletes are surprising and extraordinary. The entire system of sports media professionals works to allow spectators to come closer to understanding what such achievements entail and how they feel, and this trajectory leads inevitably in the direction I describe above. The first steps toward get-

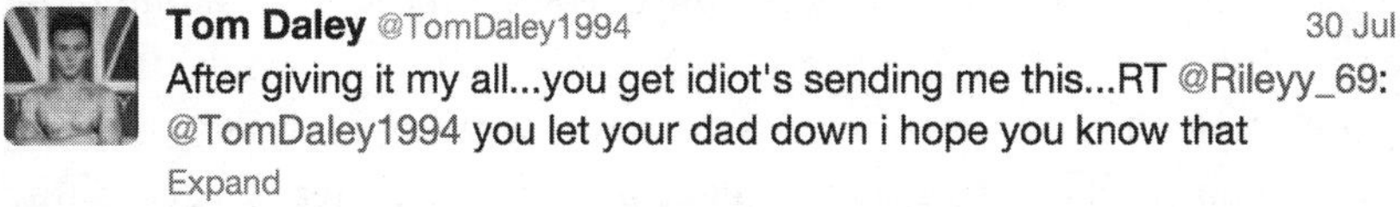

Tom Daley @TomDaley1994   30 Jul
After giving it my all...you get idiot's sending me this...RT @Rileyy_69:
@TomDaley1994 you let your dad down i hope you know that
Expand

Figure 3.1
Tom Daley's tweet.

ting inside the athlete's mind and body during competition can be found within present-day social media.

The impact of social media on athletes was most visible during preparations for the 2012 London Games, when a number of athletes found themselves in difficult circumstances because of something that had taken place on Twitter. Just before the Games, the Greek athlete Voula Papachristou was excluded from the London 2012 team before even arriving to the city because of a tweet that the Greek authorities deemed to be racist.

Once the 2012 London Games were underway, some athletes suffered because of abusive comments made by users of social media. For example, in an especially cruel tweet, the British diver Tom Daley was abused after failing to perform as well as his fans expected. (See figure 3.1.) Of particular interest in this case is that the abusive tweet probably wouldn't have generated as much debate if Daley hadn't re-tweeted it. This was the primary cause of a public debate about the abusive tweeter and a broader discussion about whether athletes should be using social media at all around competition time. Had Daley been advised to ignore the comment, there might not have been such a debate, the 17-year-old Twitter troll might not have been arrested, and the troll might not have gained 100,000 followers on Twitter—more followers than some of the most famous athletes in the world have. (The troll's account was eventually suspended indefinitely.)

The Olympic Games—or any major tournament—are always likely to be a primary mechanism for promoting an athlete's celebrity status, which may translate into sponsorship deals that can aid the athlete's training. The example of Nick Symmonds may be the most eloquent articulation of a possible future in which athletes literally sell their body parts off to the highest bidder in order to compete professionally. In this respect, the impact of social media on an athlete's profession derives crucially from the greater expectations of sports fan for real-time contact with their athletic heroes. Of course, this is one of the core values of social media as well; it has the capacity to change the structure of communication to permit more direct, intimate conversations, in comparison with older formats of communication. This shift also brings a burden of growing expectations: Some athletes may feel overwhelmed by demands from fans. Yet it is also a way for athletes to take more ownership of their celebrity status as increasing numbers of journalists rely on direct quotations from social-media platforms rather than sourcing their own unique words to report.

An elite athlete's performance space is dramatically reconfigured as a result of digital technology. Athletes now prepare and compete within environments that are awash with data systems that serve to communicate insights and opportunities for them to augment either their performance or their persona. Sports may risk over reliance on these digital systems, if they neglect to consider supporting the importance of obtaining insights or interacting with people. Yet such technologies as GLT are here to stay, and athletes must adapt to the new conditions. For an athlete to ignore social media today would be like ignoring television thirty years ago or email in 1997. Equally, failing to capitalize on data-analysis software may mean losing an edge over a competitor, which might be the difference between excellence and mediocrity. In each case, the conditions of fairness in sports are affected as a result of digital innovation, and these trends reveal a growing contest among technologists within this area of research. However, the conditions of elite sport may have been this way for quite some time now; digital knowledge technologies are just the latest example of how science and technology are defining what it means to operate within an elite sports world.

4   The Serious Gamer as Elite Athlete

As digital technology has brought elite athletes closer to amateur athletes and to fans, it has also brought amateur athletes closer to the world of the elite athlete. Whereas once sophisticated performance monitoring devices were available only to elite athletes with extensive resources and experts around them to interpret the data, today's amateur runner or cyclist can track his or her performance with a wearable digital band or a mobile phone. Serious amateurs have long used new technology to monitor or improve their performances. What distinguishes the digital era is the rise of the "quantified self" (Wolf 2010)—an idea that has recently been popularized to describe the manner in which people use digital bio-monitoring devices to capture information about their biological well-being, whether through exercise, food consumption, or simply sleep. These features are integral to the growth of interest in our increasingly data-driven society and the rise of consumer products that automate the interpretation of that data into simple messages about one's health and how one may be progressing toward life goals. Though much has changed in a very short time, the origins of these aspirations can be traced back nearly fifty years with the emergence of exercise machines.

Some of the earliest ergometers were designed to isolate certain movements with a view to optimizing amateurs' training for sports. For example, William Straub's development of the home-based running treadmill in the 1960s ushered in a new era of domestic exercise machines. What distinguishes the digital exercise devices of the past decade from the analog exercise devices of the 1960s is that digital technology has pushed performance tracking out of the home and into the public sphere. Not long ago, gymnasiums were being developed with similarly sophisticated measuring tools, such as heart-rate monitors and step gauges, built into them. Now such tools—and many more—are available as mobile phone applications, allowing people to re-think where they exercise and what they are doing while being physically active.

Mobile devices deliver data insights into the biological changes a person experiences through physical activity, and this chapter focuses on the many ways in which this is occurring. It observes that physical activity is becoming increasingly mediated through digital technologies, notably through the *gamification* of exercise and physical activity. This starting point provides a context for a wider inquiry into the relationship between digital gaming and sports, building on the conceptual origins of chapters 1 and 2. The gamification of exercise by digital technology (for example, the application Zombies, Run![1]) edge humanity toward a complete synergy between exercise and e-sports, or sports gaming. There are a number of reasons I offer in this book for why it is crucial for sports organizers to look at e-sport as a measure of its future, and this is one of them.

This chapter considers the implications of digital technology for the amateur athlete as a series of gaming experiences. In this respect, it employs the definition of *gamework* by Ruggill et al. (2004) to characterize the scope of the cultural analysis of the gaming medium. It describes the rise of computer game technology in gyms, the increasing physicality of home game console interfaces, the rise of mobile monitoring devices, and the broader emergence of *exergaming* as the mechanism through which recreational sports experiences are gamified. This development explain how physical activity is getting closer to digital gaming, but also how the reverse is also true: Digital gaming is becoming more like sport. For instance, at the 2016 Summer Olympics the arcade version of Mario and Sonica at the Rio 2016 Olympic Games—a game that requires players to run, jump, and move their arms as they compete in virtual Olympic sports—was available in the entertainment center of the athletes' village. Consequently, this chapter also explores how amateur digital sports games demand remarkable degrees of time investment and skill development, which reveals the emergence of a new sports subculture: the e-sports movement. Indeed, to demonstrate the complete reversal of the unreality hypothesis—that a game is a simulation of reality—Ruggill et al. note that some simulation games are the *primary* means of training people for the real world, flight simulators being a common and widely used example:

Flight games, for example, represent the most complex computer-game genre, often requiring players to master more than a hundred controls. Knowledge of three axes

of motion; speed controls; take-off, flight, and landing procedures; tracking and eva-
sion protocols; weapon selection; rules of engagement; and mission prioritization
are de rigueur in titles such as Jane's ATF and Wing Commander V—Prophecy. The
instruction books for such games are often intimidatingly thick, the fan communities small, and their members detail-oriented. ... In games such as these, players
must not only work to understand game play—what must be done in order to win—
but also work to understand the complex technologies and rule systems by which
a game's objectives are to be met. Flight games and other simulations actually draw
on pre-existing subcultures to form new ones, and these new subcultures in turn encourage new industry developments, new jobs, and new appreciations for the people
who do the real work these games simulate. (ibid., p. 302)

This is an eloquent articulation of how transference between spaces of
play and work spaces occurs but it also acknowledges that serious gaming
requires a work ethic that underpins play—an ethic similar to that of elite
sport. When we examine these ethics in more detail, they reveal even richer
similarities.

In pursuing this line of inquiry, I articulate how the development of
digital technologies can foster new forms of sport participation and community. Furthermore, I argue that new trends in physical digital culture
are transforming computer culture and, as a result, challenging entrenched
negative preconceptions of computer game playing as simply promoting
sedentary lifestyles or anti-social behavior. Against these hypotheses, I
argue that new forms of sports gaming make it possible for computer game
playing to support more active lifestyles. I also describe how the term *amateur athlete* cannot be straightforwardly applied to discussions about digital
interaction and sport, since the idea that, say, a computer game player is
either an amateur or an athlete is not easily distinguished. Indeed, as will
be discussed in later sections, one of the implications of digital culture is
the further conflation of the distinction between work and leisure, which
the professional/amateur divide typically describes (Miah 2011). Thus, I
explore stories of amateur participation in sports, in which there are varying
degrees of athleticism. The broad implication of this case is the consequent
inability to continue thinking about amateur athletes as isolated from elite
athletes, as they are today. To the extent that sports competition is a career
choice involving contracts, salaries, and prestige, the emerging e-sport athlete is fast beginning to occupy this space in equal measure. These ideas
engage theoretical research in leisure and play theory, but also have con-
crete implications for how sports industries must think about their communities and the means by which they engage them to develop their sports
in the future.

Computer Games in Context

Over the past thirty years, studies of computer game playing have been drawn into media debates about whether or not gaming is psychologically damaging for the players, notably the younger generation. In this respect, computer games have been treated in the same way as new formats in film and in music have been treated in the past: as potentially destabilizing cultural artifacts. These anxieties have been exacerbated as more technology has found its way into the hands and bedrooms of children and as digital systems and devices have become ever more pervasive (Bovill and Livingstone 2001). Furthermore, such worries have, in the case of other media, been subsumed within a broader set of "moral panics" (Cohen 1973) about media consumption. In the case of such moral panics about digital media and living virtual lives, Chambers (2012) notes how they "persist about socially disengaged young people replacing human contact with virtual contact." These competing views on the effects of media go back much further than digital games. Throughout history, novel cultural pursuits have been placed under public scrutiny, particularly when they have emerged from youth countercultures. The Goth, skateboarding, and surfing subcultures are all imbued with some anti-establishment characteristics, which are used to question (and often subvert) their legitimacy by controlling social structures (for example, the use of anti-skateboarding features in urban architecture).

In various parts of the world, these anxieties gained momentum as government inquiries drew further attention to the speculative and unknown effects of media culture, which elevated doubt to the level of risk. For example, in the United Kingdom the popular psychologist Tanya Byron and the Oxford University neuroscientist Susan Greenfield have publicly asserted that young people should relinquish their digital devices and play outside instead, often receiving criticism from media scholars who question the scientific credibility of their concerns (Levy 2012). The "Byron Review" (Byron 2008) notes further that "children's use of the Internet and video games has been seen by some as directly linked to violent and destructive behavior in the young," though it is cautious not to presume that this is a causal relationship. There is a long history of scholarly reactions to such claims, and there is a similarly long history of rejection of these claims by media scholars.

As a relatively young media form, computer game playing has been subjected to similar accusations. Indeed, it has been the most popular scapegoat for isolated violent events that have occurred in the past fifteen years. In the cases of the 1999 Columbine High School killings and the 2013 trial of Christopher Harris (in which the defense cited Harris' exposure to violent video games), the corrupting force of game playing has been offered as an explanation for wrongs done by young and otherwise innocent people. In the Harris case, a psychologist was called by the defense as an expert witness to support the computer games argument, though the merit of his testimony was undermined when the prosecution asked him whether the game *Pac-Man* (in which the protagonist eats ghosts) could be considered violent and the psychologist said that it could (Rushton 2013).

Computer games have historically also been criticized as inherently antisocial practices that promote sedentary lifestyles and result in diminished social skills and a lack of participation in supposedly richer pastimes. Yet, as for other media forms, evidence to support these claims is limited—particularly evidence to support the more serious concerns that are advanced about such media participation. Vastag (2004) notes that a review of psychological studies of "video game violence" identifies, at worst, temporary modifications, rather than modification of traits. Furthermore, the isolation of gaming as the culprit is a difficult position to sustain. For instance, in the depictions of the Columbine incident in two feature films— *Bowling for Columbine* (2002) and *Elephant* (2003)—computer games are present but other things that may have shaped the perpetrator's lives are also depicted as having been a part of the children's lives. In *Elephant*, viewers are shown the young perpetrators playing piano and drawing, two activities the film subtly presents as constitutive features of their leisure experiences.[2] *Elephant* urges the viewer to question the attention given to gaming as a crucial determinant of violent behavior. Even in the cases of games that have explicitly anti-social content, such as *Grand Theft Auto*, connecting game-playing behavior with real-world behavior is, at best, tenuous as a causal explanation.

There are facets of this tension between advocates and critics of computer game culture that shed light on why it attracts such controversy. After all, computer game playing remains understood mainly by young people. Though an increasing number of adults play computer games (those who were playing *Pong* in the 1970s are likely to be in their fifties now), it has always been primarily a youth subculture. It is a pursuit that is mostly not understood by adults or policy makers, and that even excludes such people. In this respect, there is an inherently antagonistic facet to the problem of understanding what happens in computer game environments, which may explain why many such youth culture transitional leisure pursuits have attracted similar discourses.

One of the problems with isolating gaming as a cause of anti-social behavior is that it neglects to consider the full range of digital game-playing activity that takes place. Though it is often the most violent games that are discussed in the news media, computer games are much more diverse in what they require of players, making it near impossible to identify one kind of game playing, as a catch all term for such experiences.

The impact of this history on the present debate about sport is analogous to how the war on drugs in society has shaped—for the worse—the anti-doping movement in sport. Thus, anxieties about broader societal problems have an impact on sport policy makers and, while there is no background of resistance toward the integration of gaming within sports, the default presumption is that these are not natural bedfellows. However, in the case of computer games at least, all of this is about to change and the beginnings of it are found in the e-sport movement. Further evidence of how this transition will occur are found by also looking into narratives around the need for young people to be more physically active and the rise of mobile health gaming experiences.

Dance Dance Revolution and the Birth of Games for Health

In the past twenty years, a number of gaming experiences have begun to challenge some of the stereotypical assumptions about what gaming requires of players and what kinds of consequences it might have for sociability or physical activity. A good case in point is the game *Dance Dance Revolution* (*DDR*), which was launched as an arcade game in 1998. Soon after its launch, *DDR* tournaments sprung up around the world and began to reveal interesting new configurations of computer-game-playing subculture. Video interviews of young players at a competition in the United States by ethnographic researchers from the University of Chicago's Video Game and Cultural Policy project revealed how the game broke through existing barriers to physical activity in young populations and created mixed-gender spaces of physical activity that were social at their core. Subsequent research by Jacob Smith (2004) found that new communities were emerging around the game. These communities had different gender ratios than traditional computer games and sports communities. *DDR* was able to attract 50 percent female players, compared to 16 percent for other kinds of games available at the time. Through *DDR*, some of the major assumptions made by critics of computer game culture were being challenged. Further research has revealed the fragility of these assumptions. For example, Crawford (2005) finds that active game players are also often physically active people, and Fromme (2003) finds that sport participation is not influenced by digital gaming habits. So, as it turns out, not all computer game players lead sedentary lives, stuck behind screens in dark rooms.

New kinds of digital gaming experiences are diversifying what was once assumed to be a monolithic culture. A good example of this is found in *pervasive games*, which are best described as mixed-reality experiences in that players participate both when they are behind a computer screen and when they are out on the streets (but still behind the screen of a hand-held computer). Describing an example of immersive gaming that demonstrates these blurring spaces, McGonigal (2003) explores a virtual community called Cloudmakers, who were inspired to act after the September 11 attacks on the World Trade Center. McGonigal explains how these gamers were empowered by their problem-solving capacities in game environments and how their example is now one of many in which game players are not sure whether the problems they are solving are real or unreal. In Blast Theory's pervasive game *Uncle Roy All Around You*, players are "equipped with hand-held computers and wireless networking, journeyed through the streets of the city in search of an elusive character called Uncle Roy, while online players journey[ing] through a parallel 3D model of the city [were] able to track their progress and could communicate with them in order to help or hinder them" (Flintham et al. 2003, p.168).

The integration of digital and physical interactions reveals how gaming is becoming increasingly demanding physically and more like a sport. This is not to say that a requirement of being designated a sport should be relative to the amount of physical exertion required of the athlete. But insofar as the amount of physical exertion may be one relevant factor, it is useful to note how it is changing in digital game playing. Even if the communities devoted to such novel forms of game playing are now smaller than those devoted to the top-selling games, which tend to involve only indoor game playing, these new experiences are growing quickly and the trajectory toward greater simulation leads to further integration of such physical exertion.

Extending the idea of gaming as a social form, consoles such as the Nintendo Wii have transformed computer game playing into a physically demanding activity. This changes the expectations one must have of game playing and the kinds of people who one would expect to play. The Nintendo Wii franchise is the fourth-highest-grossing video game franchise in the world, after Mario, Super Mario, and Pokemon (Wikipedia 2015). Nintendo Wii has also brought with it a shift in how families interact around gaming, moving the consoles from the bedroom to the living room (Chambers 2012)—a phenomenon assisted by the emergence of large flat-screen televisions, which tend to be located in living rooms and which can provide extraordinary, immersive gaming experiences.

In various bodies of literature, active gaming is referred to as "ExerGaming" (Kamal 2011), though Millington (2014) criticizes this characterization for its neglect of how such technologies exert a disciplinary force on people, compelling them to comply with government agendas and social norms around healthy behavior. Instead, Millington draws on Foucault to introduce the concept of "Bio-Play" and "Bio-Games" to emphasize these dimensions of the digital healthy games experience. Rich and Miah (2009) also highlight the theme of surveillance that is apparent within digital games for health. As part of the culture of statistics that surrounds digital environments, gaming for health—often through mobile health (mHealth) applications—has incorporated this dimension.

Today, people use mobile phone apps such as Run Keeper to track their physical activity, but also to share what they have done across social-media networks such as Facebook and Twitter. What was previously mostly an individual activity for many people has become a gamified, social experience through mobile technology. This act of using devices to share performance achievements should be seen with caution, as it is not simply about celebration and sharing. Rather, the act of publishing bio-play data exerts a disciplinary effect on players by holding them accountable to the public gaze and any subsequent evaluations the public may make upon the conduct of our lives. Nevertheless, the new wave of computer games (e.g., Eye Toy, Eye Toy Kinetic, GameBike, Nintendo Wii) offer full-body immersion and greater interactivity, providing new ways to enjoy a healthy, active life-style. They are a new chapter in how society appraises the worth of digital gaming experiences.

The Growth of the E-Sports Industry

The rise of digital gaming culture more widely is also a significant cultural phenomenon, which has an almost entirely distinct economic value com-pared with wider computer culture. The 2014 Global Games Market Report valued the worth of the industry at $81.5bn, more than double the size of the film industry (Wikipedia 2015). The total estimated revenue for 2016 is $99.16 billion, up 8.5 percent from 2015, and 37 percent of sales are on mobile devices (Newzoo 2016). The range of computer game cultures is vast, spanning a number of genres from first person shooting games to strategy and role-play games. Kuhn (2009) cites five categories of electronic games: "console/handheld, wireless, online, PC, and arcade." This use-fully reveals the range of spaces that gaming occupies from the household to the high street and even the gymnasium. For many years, the gaming market was dominated by three major manufacturers, which, according to Kuhn (2009), "account for 80%" of all game revenue: Sony, Microsoft, and Nintendo. The demographic targets of these devices vary: Sony targets 16–24-year-olds, Microsoft 18–34-year-olds, and Nintendo 8–18-year-olds. Sony's dominant share of 69 percent includes "having sold more than 120 million PlayStation2 units worldwide and more than a billion software units from its library of over 1,500 titles" by October 2007 (Kuhn 2009, p. 258). This is beginning to change, but some key characteristics of the offer by gaming companies remains the same. For instance, the prices of game consoles have remained relatively low—indeed, below cost—to pro-vide a platform for software units to reap the profits. Kuhn also reveals that, despite massive growth in recent years, the mobile gaming market was expected to remain smaller, while online games are the fastest growing seg-ment of online entertainment (including such games as SecondLife, *World of Warcraft*, DOTA2, League of Legends, and Starcraft II). It is also worth noting that, almost a decade ago electronic games represented "one-third of the toy industry in the US" and that "50–60% of all Americans over the age of six now play, with an average of 35 yrs" (Kuhn 2009, p. 260). Today many toys are considered to be cross-over games. A good example is the

Hot Wheels toy car franchise, which is accompanied by a playable mobile game. A user can purchase a physical toy car and then use a QR code to scan the purchase into the mobile game and play with the new toy car within the mobile game. Kuhn notes that, although the gamer demographic tends to be "males in their twenties," the Entertainment Software Association estimates that 41 percent of gamers are women or girls. The motivations for gaming are also interesting to observe, as they bring into question our assumptions about the characteristics of this subculture. Thus, "challenge and competition, enjoyment, social aspects" (ibid., pp. 261–262) all feature in the reasons why players seek games, which may not differ much from what motivates people to get involved in sports. Chambers (2012, p. 77) also draws attention to how the rise of digital games corresponds with the changing family in the 21st century:

Social gaming such as Nintendo Wii corresponds with transformations in family re-lations, home-based leisure and images of the digital game industry. Radical changes in family life coincide with changing forms of communication technology, social ties and spatial configurations in the home.

Sports-related computer games continue to be among the best-selling titles each year, giving rise to impressive digitally mediated communities (Kuhn 2009) and new forms of media practice (Marik 2013). A comprehensive report on the e-sport industry published by SuperData (2015) revealed some key insights into its growth. For instance, it recognizes the e-sport market to be worth $748 million, likely to rise to $1.9 billion by 2018. It also notes the dominance of so-called first-person shooter games, and a growing prize pool in competitive e-sports of around $41 million. The report also draws attention to the creation of dedicated e-sport arenas around the world and the ability of competitions to fill such venues as Madison Square Garden with large audiences.

Despite the growth of sports computer games, there has been little analysis of this within sport studies, media studies, or cultural studies. However, there are some exceptions. For instance, in 1995 Dennis Hemphill introduced the notion of cybersport to the literature. At that time, studies were beginning to explore the role of virtual-reality applications in sport. Seven years later, my article "Immersion and Abstraction in Virtual Sport" appeared as one of three chapters on digital gaming and sport in the volume *Sport Technology: History, Philosophy and Policy*. Richard Lomax (2006) published a history of the highly successful "fantasy sports" genre, the roots of which extent back about fifty years to "game boards, player cards, dice, and/or markers," which recalls my emphasis in chapter 1 on the connections between sports and digital experiences and game playing as a broader aspect of human life.

In the past ten years, the growth of e-sports has been significant, as has the growth of research into this field (Consalvo and Mitgutsch 2013; Crawford 2005; Crawford and Gosling 2009; Taylor 2012). In 2000, the World Cyber Games were established, sponsored by Samsung Electronics and Microsoft. At their peak in 2008, the World Cyber Games brought together 800 participants from 78 countries, and in 2009 the prize money amounted to $500,000. Because it was the first major event at which sports digital gaming took place, the importance of the World Cyber Games in the development of this subculture is beyond question (Taylor 2012). Moreover, the fact that one of the main sponsors—Samsung—was also a Worldwide Olympic Partner raises interesting questions about the crossover between these two types of sports events. Indeed, in 2014 another longtime Olympic partner, Coca-Cola, came on board as an e-sport sponsor of the Riot Games League of Legends, and its role in e-sports is set to grow (Gaudiosi 2015). The alignment of these Olympic stakeholders with the e-sport industry may be no coincidence. Hutchins' (2008, p. 858) overview of the WCG values shows clear common ground with the Olympic movement:

The WCG slogan, "Beyond the Game," means that the WCG is not just a world game tournament, but also combines the world to create harmony and enjoyment through shared emotions. Further, the WCG slogan hopes for peace to emerge during its tournament, which fosters mutual respect amongst all participants from all over the world as we strive together to build an attractive "World Cultural Festival."

The World Cyber Games came to an end in 2014 when the president announced there would be no further tournaments. Though they were informally referred to as the "e-sports Olympics," there was actually only one sports game within the competition: FIFA's soccer game. In an interview published on the website Tech in Asia (Custer 2014), the WCG organizer Lin Yuxin rejected claims that the WCG was brought to an end because of increased competition from other events or because of the failed move toward mobile gaming. Instead, Yuxin identifies the shift in Samsung's strategy toward mobile gaming as crucial, since the WCG was a PC gaming competition. Yuxin also argues that the Olympic style of the WCG was in excess of its brand value, leading to over-commitments and excessive administration. Finally, Yuxin notes that the WCG failed to capitalize on its "golden age," and that losing the interest of Samsung was catastrophic to its future.

Yet the WCG was not the only games competition out there, nor was it the only competition to cease doing business. In 2008, the *Championship Gaming Series* founded by DirectTV came to an end after just two seasons. The Electronic Sports World Cup, which featured Konami's Pro Evolution Soccer in the years 2004–2007 and has featured FIFA soccer since 2009, has been more successful. With the slogan "e-Sport for all," the ESWC has brought together competitors from more than 100 nations and has offered more than $2 million in prize money since its inception. Also, since 2008, e-sports has had its very own federation—the International e-Sports Federation (IeSF)—which has set out to champion e-sport as a legitimate sports industry. The importance of this should not be overlooked, as the emergence of a dedicated federation may be akin to the codification of traditional sports that took place in the 19th century. Bringing together an increasingly fragmented and commercially diverse community is no minor challenge. Nevertheless, important achievements have taken place in this direction. Since 2013 the IeSF has become an official signatory of the World Anti-Doping Agency, and in 2014 it became a temporary member of the Sport Accord, the first step toward being fully ratified as an International Sports Federation. The IeSF e-Sport World Championship (not to be confused with the electronic Sports World Cup) took place under that title for the first time in 2014, having previously been called the IESF World Championships. FIFA soccer was a major feature of that tournament, and baseball, racing, and combat sports all have made appearances on the schedule since the inaugural tournament in 2009. Furthermore, the IeSF has managed to win recognition as a sport from more than fifteen national governments in its short history and is also working with other sports federations. For instance, in 2014, the IeSF and the IAAF launched a joint program with the slogan Athletics for a Better World, which set in place a basis for their continued collaboration. And in May 2016 the World E-Sports Association was launched by the Electronic Sports League, which aims to establish professional standards across the industry.

Despite these events and organizations, there is an absence of games with explicitly sports-related content within these international tournaments.

For example, the three titles played at the 2015 IeSF World Championships were *League of Legends, Hearthstone,* and *Starcraft 2,* games clearly located in the fantasy genre. The absence of sports related titles is explained by the relative numbers of gamer communities or the challenges associated with developing worldwide tournaments of this kind that focus only on sports. It might also reveal the nature of e-sports team structures, which tend not to be restricted to just the sports genre. Nevertheless, for now, the term *e-sports* is used as a catchall term, into which any form of competitive digital gaming is being included, whether or not the game has any sports-related content. This may prove to be one of the limitations in achieving full ascendance into the world of traditional sport and further fracturing may ensue in coming years.

With the creator of *World of Warcraft* saying that it is time for computer games to be given sport status (BBC 2014), it may just be a matter of time before they are brought into the fold of traditional sports. However, if the games played at major tournaments are more like military games than sports games, it is difficult to see this leading to any kind of ascendance into traditional sports territory. Moreover, to the extent that military titles rather than sports titles are the most popular, it could be a financial disaster for the games industry to walk away from those other titles and focus on sports. Indeed, examining the lists of the most prominent international e-sports tournaments in the world makes it apparent that games such as *League of Legends* and *Defense of the Ancients (DOTA)* are the most prominent kind of game (e-Sports Earnings 2015). Thus, playing fast and loose with the concept of sport may be a strategic necessity at this point in the history of e-sports. Nevertheless, discussions are beginning to take place about making e-sports fit more widely into the traditional structures that describe traditional sports. The media coverage of such communities has also recently provoked questions as to the relationship between these sport-like communities and more traditional sports, with the BBC asking "Are pro video-game players our 21st century athletes?" (2012) and "Is Computer Gaming Really Sport?" (2015).

With the longtime Olympic sponsor Coca-Cola sponsoring e-sports since 2001 and the emergence of online streaming platform Twitch broadcasting tournaments live, there are many questions about how e-sports may adopt or destabilize traditional forms of monetizing broadcast content. The Twitch model of monetizing broadcast content may mean that e-sports bypass television deals altogether. Moreover, this signals a longer-term shift in sports broadcasting more generally as everything moves to online only. Alternative, the in-game micro-transactional economy of e-sports gaming may offer a completely new model for monetizing digital media experiences.

With an average age of 21 years for Twitch viewers and an average age of about 54 years for American television viewers, there is a huge disparity between these audiences (Thompson 2014). Moreover, since Twitch was purchased by the leading online retailer Amazon for just under $970 million in 2014, there is even more reason to conclude that the online-only model for distributing gaming content will span far into the future. However, in the year 2014 ESPN broadcast the finals of The International—the major *DOTA 2* tournament—despite the fact that its president, John Skipper, said in the same year that e-sports are "not a sport" but rather "a competition, more like chess and checkers than 'real sports'" (Skipper, cited in IGN 2014).

e-Sports Athletes as a New Elite

Cyber-athletic competition cannot be thought of in terms of media or sport or computer gaming. The institutional and material boundaries separating them have imploded, leading to the creation of a new social form, e-Sport.
Brett Hutchins (2008, p. 865)

As e-sports have become more professionalized, there has been greater recognition of the participants' achievements. For example, in 2013 e-sports athletes gained formal recognition from the US Immigration and Citizenship Services, which decided to assign professional e-sports players the same visa status as other athletes (Robertson 2013). An important part of the professionalization of e-sports is its integration with broader elements of the entertainment industry, specifically around the licensing of computer games. Though the prefix 'e' in the quotation from Hutchins and in the wider world of e-sports may not stand the test of time (as is also true of countless other Internet-isms that have come and gone from digital parlance), the 'cyber' lexicon serves as a way of emphasizing an important cultural and economic shift within society that is brought about by digital technology.

In sports, perhaps the one dimension that connects the elite athlete with the amateur athlete, beyond their competing in a similarly designed arena and with similar apparatus, is the area of computer gaming. Gaming culture has already given rise to subcultures of cyberathletes in which computer game players have celebrity status as athletes, including their own sponsorship deals. For example, the renowned cyberathete Fatal1ty (Jonathan Wendell) won more than $450,000 in prize money in the years 1999–2006 (e-Sports Earnings 2015). In the period 2009–2014, Jiao "Banana" Wang earned more than $1.8 million from e-sports (ibid.). Since 1998, the total prize money for e-sports tournaments has grown considerably. (See table 4.1.)

As e-sports become more professionalized, it becomes increasingly difficult for critics to deny that playing computer games requires the same

Table 4.1

	Total prize money	Average tournament prize pool	Total tournaments	Average earnings/ player	Total active players
2015	$65,017,976	$14,630	4,444	$5,184	12,542
2014	$35,744,202	$18,763	1,905	$6,182	5,782
2013	$19,472,753	$14,172	1,374	$4,852	4,013
2012	$12,963,274	$9,293	1,395	$4,068	3,187
2011	$9,570,126	$8,583	1,115	$4,138	2,313
2010	$5,089,479	$8,289	614	$2,701	1,884
2009	$3,346,809	$10,459	320	$2,588	1,293
2008	$5,964,363	$20,218	295	$5,870	1,016
2007	$5,732,693	$23,986	239	$5,626	1,019
2006	$4,297,790	$16,723	257	$4,109	1,046
2005	$3,529,215	$15,147	233	$4,024	877
2004	$2,057,645	$13,903	148	$3,287	626
2003	$1,309,012	$16,570	79	$3,348	391
[2002	$833,606	$27,787	30	$2,053	406
2001	$772,573	$24,922	31	$2,782	269
2000	$514,078	$19,040	27	$2,706	190
1999	$245,971	$20,498	12	$5,466	45
1998	$89,400	$17,880	5	$4063	22

Source: e-Sports Earnings 2015

kind of skills and perseverance that any athletic endeavor would. Even the absence of gross motor activity does not go very far in discounting e-sports activity as un-sport-like, since athletes train their bodies in ways that are analogous to what, say, a racing-car driver does in preparation for his athletic performance, which involves mostly sitting. Like drivers of racing cars, e-sport athletes sit down for their competition, but each has to be in a state of high fitness in order to be competitive.

One of the most interesting dimensions of the integration of virtual and physical worlds is the way the real-world career of an athlete or a team has blurred with the in-game narrative created by game developers (Silbermann 2009). This is most obvious in the player-manager games; for instance in FIFA's soccer game, the avatars for players are created with the skills and performance characteristics of the real players from the most recent season. This layering of fantasy and real-world contexts is a further indication of how the worlds are becoming interdependent. Indeed, Silbermann notes that "soccer video games were successful in helping athletes have a better understanding of various styles of play" (ibid., p. 169).

Through computer games, the career of an elite athlete becomes part of a new narrative, which, in turn, extends the fictive space of sports to a new territory in the pursuit of leisure. It is even apparent that data on players uploaded to games feeds into the analytic discussions that surround actual players and their recruitment. For example, the simulation game Football Manager works with ProZone Sports to extract data from within the computer game environment and integrate it with the professional recruitment database owned by ProZone. Stuart (2014) explains this as follows:

To ensure authenticity, Sports Interactive has spent the last 22 years building its own network of "scouts," dedicated Football Manager fans who attend real-life matches and training sessions, and then file detailed reports on players so that the game's database is authentic. Some scouts watch a single team, others a whole league, and all remain in regular contact with the development studio while swapping tips and experiences on the company's buzzing forums (online).

This is not the first time that there has been a transfer from a game environment to actual sport. In 2012, Football Manager also created a real-life manager "when Azerbaijani student Vuagar Huseynzade was put in charge of FC Baku's reserve team based on his success in the game" (Rumsby 2014).

All these transformations to e-sports reveal that there is nothing inherent in traditional sports that permits their being afforded a special status as physical cultural pursuits that should be distinct from computer-based sports games. As the conditions of the gaming industry align with the values of the elite sports world, and as the overlap between these two imaginary worlds of performance grows, we may yet see a world in which computer gaming is given serious consideration as a future Olympic sport. Indeed, at the e-Sport World Championships in 2015, representatives from the International Olympic Committee were in attendance, and so there may be a will to understand how the world of sport is changing toward gaming from the most ancient of sporting institutions. Indeed, if BMX biking, mountain biking and surfing can make it into the Olympic Games, then why not gaming? The economic argument alone may see sponsors quickly shifting their investments into the growing market of e-sports gaming and away from traditional sports events, unless the latter begin to adapt and embrace their inevitable digital futures.

Yet there are also deeper, ideological reasons for why e-sport may be a sign of the times for traditional formats. Consider how the modern Olympic Games revival was predicated, in part, on a desire to bring about social change. Today, one of the key aspects of such change is the ability to be active in digital environments. Indeed, one might even argue that Internet access should be a matter of human rights, or might think of it as a public utility comparable to electricity or water. As more of civil society moves into online space, there comes a point at which failure to access information may limit one's democratic rights. In this respect, e-sports should be seen as a vehicle toward creating a more digitally literate society, a crucial aspect of ensuring that people are optimally placed to respond to some of today's most important global issues, one significant element of which is about getting the entire world's population online. In this sense, advocates in e-sport can peg their social mission on the basis of promoting digital inclusion and access to digital skills development. After all, if a nation is to become competitive in e-sports, it will have to develop a globally competitive digital infrastructure. At a time when architects are building the world's first e-sport stadiums it is worth considering that what happens in such environments need not just be competitions. Instead, the future arena may include various kinds of exciting learning initiatives, designed to solve society's biggest problems. This way of imagining sport's future would be appealing from the perspective of sustainability, but also would allow sports to appeal to the importance of still having physical arenas, in an increasingly virtual world. The future of the sports competition venue is reliant on its capacity to be a focal point for much more than just sport. However, this is not the only social issue around which e-sports can seek to make a difference. At present, one of the big problems in e-sport competitions is the absence of female players. Yet instead of feeling unable to make significant changes in this area, e-sport may institute interventions that allow it to become a leader in addressing inequalities in digital competencies between genders.

For people outside of the e-sports world, a world in which digital games could become Olympic sports may seem a very unlikely scenario. Indeed, there are many hurdles to be overcome before this could ever be possible; one important one is whether the e-sport community even aspires to such a status. Even achieving a situation in which e-sports could have their own equivalent of the Olympic Games, in which a range of games would be played, is difficult to envisage, as at present only a few titles can claim to have large enough communities to stage such events, and combining theirs with others is not obviously valuable. Yet this is also what makes the future of e-sports so interesting, as, to some extent, it will be a test of the value of human solidarity to see whether e-sports embraces the more traditional sporting ethos that allows a range of federations to come together in a common endeavor. While there is more that the larger sports federations can do to support the smaller ones, there at least exists some sense of common purpose and the future of e-sport may be to discover this and realize that, games that may be quite "niche" or emerging are nevertheless crucial to support by the larger titles and publishers, so as to nurture the evolution of gaming and the diversity of its player community. Such aspirations would also allow gaming to occupy a more central and influential political role in sport and outside of it.

However, for the skeptic, it is crucial to consider that the trajectory toward e-sports is not solely in the rise of gamer communities but also through the digital gamification of traditional sports. As sports evolve, more digital elements will be brought into the gaming experience for athletes, transforming them into mixed-reality encounters. These kinds of e-sports will have a significant impact on what one considers to be either gaming and sport.

5   Digital Spectators in Augmented Realities

To conclude part II's focus on how the elite sports environment is altered by digital technology, this chapter focuses on the *spectator* experience. Changes in this area may be traced to longer trajectories of media development than simply the digital era and, over time, this shift has nurtured new ways of participating in sport as a spectator. Often the focus of these new experiences is the creation of opportunities for interaction between producers and audiences. Indeed, the fascination with interactivity in the 1990s reveals an absence in how we theorize and make sense of our relationship to mediation. After all, since the dawn of the media age—perhaps even since the earliest forms of recorded history—authors have attempted to foster an active relationship between their content and the receiver. Seo and Jung (2014) even tie the rise of e-sports gaming to the desire for additional spectator consumer experiences, reinforcing the connections across this cybersport nexus, imagining players as a kind of audience.

The mode of this interaction changes with each new medium. When one reads books, interaction occurs through the imagination we invest into the words on the page, how we re-tell the story to ourselves, the environments in which we find ourselves reading, or how we revisit the text again, creating new narratives around the same words. In the early years of newspapers, interactivity involved readers taking the time to write into the newspaper editors, responding to the content in letters pages, for instance. In the age of radio, the desire for interaction was satiated through talkback programs and chat shows, a format later developed through the rise of reality television. Indeed, the history of interactivity may be stretched so far as to lose all connection with the present focus on digital transformations, but it is important to acknowledge these antecedents, if only to recognize that the key principles of what the most cutting-edge designers of virtual realities are trying to achieve today occupy similar ambitions to many of their predecessors who made similar attempts through different means. In the end, all forms of mediated content—from books to holograms—allow users the opportunity to construct narratives that occupy their imagination and which affect their perception of one world by entering another.

In the earliest years of the Internet era, the mode of interactivity was generated from the way that *users*—no longer audiences—could explore new worlds and produce their own content on their own channels. Since 2008, sports media organizations have tried to use social media to create new ways of interacting with sports, providing additional layers to the spectator experience. For instance, during the 2012 London Paralympic Games, the first-time Games television broadcaster Channel 4 (an independent channel in the UK) used Twitter to create audience polls, whereby viewers were asked to use specific hashtags when responding to questions that were asked on the breakfast show. In this sense, the spectator experience has always been changing, which makes it especially hard to foresee which trend will bring about a categorical change in how media communication takes place. Of particular relevance today is that the additional opportunities for interactivity afforded by digital technology complement, rather than compromise traditional televisual formats. This is crucial because sports are wholly reliant on the sale of television rights to broadcasters.

Many of these earlier interactive formats remain in use and, in some cases, are relatively unchanged. Even 20th-century formats are quite similar to their original design. Radio still relies on listeners telephoning the radio station to chat with presenters and share their views. Also, television chat shows are still very similar to when they were first devised, and people can still create their own. In this sense, new layers of interactivity are added to each new media ecosystem, rather than simply replacing what has come before. Indeed, rather like evolution, with each new innovation, additional functionalities are created, or older functions are honed. Sometimes, new forms of communication are to the gradual—or radical—demise of others, but many remain in place despite the creation of new formats. After all, mathematicians still rely on blackboards as much as they rely on calculators or computers. Digital artists still develop fundamental creative skills in illustration, as grounding for their innovative creative work. Filmmakers still develop ideas and scenes on storyboards, using paper in parallel, perhaps before resorting to a digital environment. Similarly, the utilization of digital technology in sport is an additional means through which to develop experiences of sport performance and not necessarily a replacement for other systems.

Until recently, such trends have hardly had an impact on the area of sport spectatorship—what takes place during the sport itself—and sports broadcasting has been staged primarily as a theatrical performance. Today, spectators can make their own, unique spectator experience, by drawing on a range of live and mediated spaces to create a personal encounter with elite sport. The consequences of this shift to the sports broadcast or the live sports experience remain unclear, but change to ways of working is inevitable. In this context, this chapter explores how digital technology facilitates today's spectator experience, which, in turn, is also remaking the kinds of expectations spectators have, the communities they form, and the way media organizations develop sports content around an audience.

Digital Sport as a Theater of Possibilities

Early work theorizing the spectator's experience is not found within the sports literature, but is located among various bodies of theater studies, television studies, and film theory. Notably, Walser (1991) describes both the context of digital worlds and sports as crucially similar, noting that "cyberspace is fundamentally a theatrical medium" (p. 51), rather like sport. On such an understanding, sports are described as events within which persons can experience their bodies and confront their embodiment or, as Walser puts it, "sports evolved out of the human impulse to assert the self, thereby ensuring survival" (p. 52). Virtual worlds provide an environment in which such encounters with our embodiment can take place and in which such encounters cannot help but be juxtaposed to the embodiment of sports.

Virtual reality presents a new comprehension of what it means to be embodied and engaged while also being physically active. Yet the origins of the human interest in virtual reality can be traced even further back. For example, Murray Smith (1995) discusses how spectatorship is intimately connected to the age of philosophical romanticism and the birth of the novel, in which dreaming is introduced as a metaphor for the desired mental state one seeks to achieve within literary encounters. The loss of awareness, suspension of disbelief, and acceptance of relocating oneself mentally in another place allows the "spectator as a dreamer" (p. 113) to emerge. While the experiences of film, theater, music, and sport audiences are undoubtedly different in many ways, they each involve a willingness to believe "that the fiction is in fact not a fiction, but either reality itself or a representation of real events and persons" (p. 114). In this sense, the fiction of sports is borne out of its gratuitous logic, dramaturgy, and its being a performance—the meaning of which Hemphill (1995, p. 48) describes as being co-produced by "actors and audiences." Indeed, Hemphill goes on to note sport's role in the civilizing process (Elias and Dunning 1992), which is intimately connected to the privatization of leisure spaces, such as theaters and playing fields.

As for other kinds of performance, the way that sports are mediated leads us to renegotiate our relationship to the live experience. Moreover, the historical development of sport spectator experiences evidences the claim that a mediated experience is becoming far richer than the live version. As Morris and Nydahl (1985, p. 101) write of sports broadcasting, "television producers can design sports spectacles laced with visual surprises that present a range of dramatic experiences which the live event cannot." They also reinforce the way in which sport experiences are made and remade through innovative media content. Morris and Nydahl note "slow-motion replay not only alters our perceptions of the action it reviews but also establishes our expectations" (p. 105). These parameters make the digitally mediated sport spectator experience particularly compelling to investigate from the perspective of future media. Indeed, the integration of live replay screens within sports venues speaks to the audience's desire for an enriched live encounter and the inadequacy of the live event—free of mediation—to provide this for audiences. While one might dispute the value of simply replaying a goal or incident in one's head, the technological realization of this through replay is clearly something that audiences have grown to value and may be, therefore, better than just what our memories can do to help us relive such experiences.

There are four overlapping aspects of the digital spectator experience that are changing what it means to be *present* at a sports event. The first of these has to do with the increasingly immersive nature of the spectator experience. Second, I consider how the use of augmented reality is transforming what we see within the playing field and the perception we have of the live event as a result of this change. Third, I explore how the use of second and third screens, and mobile technology more widely, changes how, when, and with whom one undertakes spectator experiences. Finally, I examine how the use of social media is expanding the duration of the spectator experience and what kinds of things audiences do as social-media-enabled participants.

Total Immersion

In the same way that television transformed sport spectating by providing more varied opportunities for people to experience live events and by separating the physical presence of the spectator from the competition arena, virtual-reality technology marks a radical shift in the history of sports spectating. In each case, the developers are striving to create a more immersive experience for the audience. Back in the early days of the digital age, Hemphill (1995) imagined how virtual-reality head cameras could be worn by spectators, enabling total immersion spectating. The user would experience the performance of the athlete, as if they were the athlete. Some of these earliest designs bear a close resemblance to some of the newest technologies that are trying to deliver such experiences. Indeed, in 2014, the Australian Open worked with IBM and Oculus Rift in their ReturnServe project, which allowed fans to experience what it feels like to be on the receiving end of one of the fastest tennis serves in the world. According to IBM (2014),

During the event, visitors were able to step into a virtual Rod Laver Arena using the ReturnServe Oculus Rift headset and use a specially-designed motion-sensitive tennis racquet, so they could return a live serve from one of the best tennis players on earth.

After taking a swing they were able to explore the court in 3D virtual reality and gain an understanding of other ways big data is being used to change the game, from sport to business and beyond. Over 10,000 people tried the experience across multiple locations in Melbourne and Sydney.

The personalization of the audience experience is, thus, a crucial part of creating a greater sense of immersion. For many years, television has offered viewers the possibility of making personal selections of what they watch and how they watch it. However, it has never been possible to transform a player into a real-time data-capturing device, until relatively recently. Today it is possible to attach body cameras to athletes without inhibiting performance. For instance, the use of GoPro cameras or Google Glass each provide a near line-of-sight perspective of the athlete, drawing the remote spectator closer in to their experience and in-competition trials have already begun to broadcast using such devices. In January 2015, GoPro teamed up with the NHL to pilot the use of wearable cameras for players, offering new perspectives on ice hockey for fans. (See GoPro's website.) This is part of a new program delivered through a partnership between the wireless camera company VISLINK and GoPro once called the GoPro Professional Broadcast Solution and later branded as HEROCast. The start-up company Eye360exp is beginning to create augmented-reality experiences using 360° filmmaking—again with Oculus Rift goggles. Perhaps the day when sports are played in virtual arenas has arrived. Indeed, at the Rio 2016 Olympic Games, the BBC and NBC produced about 100 hours of coverage of 360° video content for VR experiences.

The goal of total immersion is inextricable from the project of creating mediated sports experiences. Sports media professionals are constantly trying to find ways of making the content they produce around sports more compelling, engaging, and immersive. Yet it is also one of the greatest challenges for technologists to achieve. While it would seem that every stage in the history of screen technology leads viewers to claim that the experience is lifelike, it is only when a new stage in the development occurs that the inadequacy of previous versions is exposed, along with how far they were from simulating reality. There is still a gap between what people experience in the physical world and what digital platforms can deliver, but the gap is getting narrower.

Philosophically, immersion into an alternate reality is an appealing goal for users, since the prospect invites us to consider, and subsequently experience, what it is like to inhabit another person's life—to live as if one is somebody else. In this sense, it challenges our existential presumptions about our lives and allows us to experience the sensory and corporeality of another person, something which otherwise may be impossible. The 2014 Oculus Rift gender experiments are a nice example here. Two participants—a man and a woman—have their physical bodies mapped onto each other, to create a sense of what it might be like to exist within the other's body. To work, this telepresence system relies on the user's cooperation with the other participant's movements. They have to work together, to become a unit, in order to experience what it is like to be the other. Together, they create what the project calls "embodied narratives," an experience of being another, through a machine:

I am no longer Aaron Souppouris. I am a woman. I am a stranger. I stare down at the mask I hold in my hands, struggling to comprehend how those hands, which are clearly not mine, are allowing me to feel its curves and cracks. As I glance at the mirror in front of me, my new lip piercing glimmers under the harsh fluorescent lights. This is not a fever dream, not a hallucination, not even a video game. This is The Machine To be Another. (Souppouris 2014)

In some respects, all forms of storytelling strive for this kind of otherly experience; the most accomplished stories provide detailed, intimate, and compelling descriptions of what it is like to be someone else, to live their lives, and to be taken outside of ourselves. The challenge for architects of virtual reality is to heighten the sense of realism that digital worlds promise and to allow the user to understand other worlds and other people as they are, rather than how they feel lived through the device they are wearing. The ideal virtual-reality system is one in which the medium is imperceptible and in which it prevents us from knowing that we are within a simulation.

Yet the project of virtual reality requires us to acknowledge that our sensory capacities may also need modification in order for them to deliver the kind of immersive experience we seek. In sports, this is especially true. After all, the speed and the complexity of the movements in sports are often too great for the human eye to appreciate in full without using some form of mediating technology. As a result, sports media have developed a number of techniques to compensate for our sensory or cognitive inadequacies as spectators. The use of slow-motion replay—a technological act of remembering what matters most—is a classic example from sport. This act of repeating critical moment in a competition performs two kinds of act for the viewer. First, it helps them see more clearly what took place but, more crucially, it informs the viewer what is important within a game. In so doing, the decision to replay something creates a hierarchy of what elements of the competition matter most—often moments that are considered to be pivotal in the overall result, such as goal scoring or injuries.[1]

An early example of more digitally-driven augmentations was the FoxTrax ice hockey puck, which was introduced by the Fox television network in 1996. The technology involved inserting a circuit board and a battery into the puck so that on television it would appear to produce a glowing trail. That innovation failed to capture the imagination of seasoned viewers, who felt that the computer-generated imagery interfered with their viewing experience. Newer hockey viewers may have gained more value from it. In any case, the FoxTrax puck was abandoned. This is a good reminder of the fragility of technologically immersive experiences developed in isolation from other narrative considerations. The experience of viewing ice hockey on television may have already been encoded for viewers to such an extent that altering its parameters would disrupt what they enjoyed about it—not simply ice hockey, but *televised* ice hockey. It is also a reminder that new kinds of sports media experience may give rise to completely new kinds of mediated spectator experiences. The naive view of sport's digital future is that technology will be steadily integrated into traditional sports. A more reasonable view is that existing sports will evolve and completely new sports will develop as a result of new technological possibilities.

In the same way that television audiences and radio audiences have different sensibilities, a virtual-reality audience is likely to have their own way of wanting to experience content. This is crucial for the sports media industry to take on board, particularly since discussions about transforming media experiences often become embroiled in anxieties about how to accommodate established media audiences, while also innovating. It doesn't make sense to create a radical new form of viewing, while losing the interests of an established audience. Moreover, traditional media audiences wouldn't necessarily shift their viewing experience into virtual reality. After all, even though 3D television has been in use for some years, neither audiences nor filmmakers have moved seamlessly into the 3D viewing experience. It may turn out that wearing augmented-reality goggles for the duration of a game is also a fad that fails to catch on as a mainstream experience. This is why it is also important to bear in mind that technological shift should be accompanied by cultural and behavioral adaptations in order for some transformation in spectators' experiences to take place.

In strictly non-technological, narrative terms, sports *commentary* is among the most important examples of how greater immersion for spectators is created. Other language-based elements may include pre-match analyses and post-match summaries. The common ground between the technological and the cultural aspects of augmenting sports spectating is *language*, specifically the creation of new kinds of language through which one makes sense of something. This can be achieved in numerous ways through sports media. For instance, the BBC's sending its news anchor Huw Edwards to the opening ceremony of the 2008 Beijing Games alerted the audience that there was something more than just sports history being made. Edwards' role in that moment gave greater historical and political gravitas to the ceremony, lending weight to the importance of the moment.

The history of technological transformations to mediated sport reveals how the spectator's experience tends toward total immersion rather than increased passivity. Audiences want to feel closer to the center of the sports action. Ultimately that means understanding how it feels and what it takes to perform as an extraordinary athlete. One step toward cultivating this insight is to operate as co-producer of the performance through interacting with media content. This is a further reason why traditional sports can learn from what is happening in e-sport. Indeed, Borowy and Jin (2013) describe the digital gamer as a participant in an *experience economy* that had been growing since the 1960s and which was apparent within the rise of gaming culture. These characteristics are particularly fitting to describe what takes place in fantasy sport (Hutchins, Rowe, and Ruddock 2009), another vastly expanded format of participation in sports (Lee, Seo, and Green 2013).

More broadly, the goal of sports media resonates with these aspirations to bring spectators closer to the action and help them understand it in its entirety. This is not to say that the spectators are always becoming more active during their experiences. After all, a spectator may not want to—or be able to—run alongside his or her favorite runner—which would presumably be the pinnacle of such achievements. Yet this alignment of spectator and athlete describes how spectators are becoming more integral to the production of sports events and, as a result, are an important component of how sports are being remade through new media. The nature of this relationship is dynamic, unfixed, and persistently unknowable, as the future of immersive experiences and technology is unknowable. We don't yet know what audiences 50 years from now will consider to be a rewarding experience or what it entails to be close to the performers.

The experience of total immersion is best imagined as a continuum on which narrative and technological elements come together to remake the experiences of sports spectators and on which perceptions of being present and fully immersed shift over time. For that reason, I am cautious to not present a given future of sports based on some present-day technological interface. Even today's emerging virtual-reality systems do not reveal much about what future sports spectators' experiences could look like. Similarly, today's assumptions about what would constitute a perfect simulation are inadequate reference points, since our perception of the simulation is limited to present-day cognitive and sensory functions and limits.

In part, our inability to imagine the future—and the broader problem of futurology—is due to the fact that so many aspects of that experience are technologically determined and technology operates within complex systems. Imagine a virtual-reality experience of golf. Virtual reality has been utilized for some years to help players improve their golf swings. From the most primitive to the most recent examples of golf simulators, camera and sensor devices are used to monitor the swing. The player is physically performing the same movement she would perform on a golf course. Once the ball has been struck, the simulator uses a screen in front of the player to show where the ball would have gone on a particular course, had it been struck in this manner. Broadly speaking, this technology was not developed for the spectator experience, but it could be. Imagine further that, as a spectator in a remote location, you could watch the player take the swing and then, once the ball is struck, your point of view switches to the ball itself, not observing the ball as it flies, but observing it from its position in the air. The ball itself becomes our eyeball and we fly with it. Add to this the use of haptic feedback, whereby spectators actually experience the sensation of the ball in flight. Together, these elements create a novel form of audience participation, which attests to the idea that our imagination of what constitutes a perfect simulation is fluid. Taking this example further, imagine that, as the ball comes back down to earth, the spectator becomes virtually separated from it, adopting an observer or third-person viewpoint, or continues on its path and feel the bumps and lumps of the fairway as the ball journeys toward the hole.

Sports broadcasters have not yet imagined this kind of spectator experience; at present it is an unknowable audience experience, perhaps combining elements of traditional viewing and aspects that are more comparable to a fairground ride. Yet it is technically not so hard to imagine—each dimple in the golf ball could be the lens of a tiny camera, and stabilizing technology could stitch together what they capture, creating a 360° spherical perspective, so spectators can move their viewpoints in whatever direction they choose as the ball flies. Yet we do not yet know whether this kind of spectator experience would appeal to golf fans. Furthermore, we don't know what kinds of sensibilities future golf spectators will have, given the wider context of media change that surrounds their lives. Imagine further that the ball's flight in the air could be affected by some kind of haptic interface controlled by an additional player. The "flight specialist" role could be an entirely new role in sports whereby we try to test the physical capacities of a player to steer a ball in flight using their own bodies.

We do not know whether the richest experience will always rely on some creative camera operator to have the skills necessary to achieve the most compelling perspective, or whether audiences will be the best people to decide the camera angle of their spectator experience. However, we do know that, in addition to creating an experience that provides a first-person view of the athlete, virtual realities create additional experiences. They do not just aim to re-create a conventional experience; they aim to extend it. In so doing, VR provides a wider range of experiences than can be achieved by just thinking about the athlete's perspective. This is a richer interpretation of total immersion than is typically offered, since it considers the athlete as only one viewpoint in a wider range of actors and props that make up a sports spectator's encounter. Moreover, approaching sport spectatorship from the perspective of relationships between space, people, and objects invites us to consider further how sports might be changed by altering the physical spaces in which competitions are held.

Augmented Reality in the Stadium

One of the most transformative ways in which digital technology is affecting the experiences of sports spectators is through the use of augmented reality. Augmented reality is still in early development within sports spectatorship but already has distinguished itself from virtual reality. Though a lot of the discussion so far has focused on virtual reality (i.e., creating simulated experiences of the physical world, either by using computer-generated images or by using camera simulations), augmented reality is distinct in that it involves adding layers of digital content to a physical-world camera perspective so as to enhance the content in some way. One early digital application that is delivering augmented-reality experiences is WordLens, which performs live translation of text seen through the camera of a mobile device.[2] The user points the device at some text and the app begins to automatically translate that text from one language to another. This effect is shown as a replacement within the camera's frame for the language that was there previously. The user's view then appears as if the physical world has the translated text written in the new language, rather than its actual form, since it also adopts the typography and colors of the physical-world text. Leaving aside the debate about whether the world is ultimately better off with live translation, rather than being populated by people who spend time learning a language, the example illustrates how augmented reality creates content that is additional to the real world and thus creates new experiences.

Applications of augmented reality are still sparse in sports, but some early examples suggest how far this technology may take the sport spectator experience in the future. For example, when pointing a mobile device camera at a basketball player, it could be possible to track his shirt color and his number, which would then cause additional information about the player to be displayed. Such information might include biographical details, playing statistics, and other kinds of information typically offered by a broadcaster during a game. These additional layers of information—and perhaps even interaction between spectators—will have huge monetary value, but will also change what spectators do when watching live.

Sports organizations need to ascertain how to adjust to these new realities. Though it may be tempting for traditionalists to retreat to the idea that just watching the sport should be enough stimulation, the reality is that, already, spectators are not just watching the games. What happens in front of the spectator in physical space is competing with a range of other information and notifications that come to them through their devices, which will divert attention. For those spectators, sports organizations will need to consider how they can ensure they provide content during the game that can ensure attention is focused on them, even if the audience is not looking at the sport itself. The challenge for sports is to re-claim their dominance in the attention economy of live events, and the best way to achieve that is to become part of the mobile experience. The data insights alone into fan behavior and trends will be possible to monetize to allow sports organizations to deliver more value for sponsors.

The transformations augmented reality will bring to physical spaces are considerable. That is why it is especially interesting to think about how a sports arena might look in a future in which augmented reality is pervasive. After all, one might presume that perfect simulations or heightened realities lead to a diminishing need for physical spaces. Yet there are reasons to be suspicious of this imagined virtual future. Indeed, to declare the end of physical space in the face of perfect simulations is what chapter 2 considers to be shortsighted, since this presumes that one can distinguish easily between physical and virtual worlds. This is not at all clear. Although some new sports may emerge that are less dependent on physical arenas, and although some traditional sports may adapt to a similar diminishing need, there is also reason to conclude that the physical spaces in which sports take place will be adapted as a result of these augmented-reality experiences, not that they will become redundant.

In a world of heightened simulations, the physical arena can reinvent itself, perhaps becoming a high-tech hub within which the most compelling augmented and virtual experiences are made available for spectators to enjoy. These new cathedrals of digital utopias would concentrate the playful experimentation with advanced technology in a way that is impossible to achieve within more domestic settings, such as living rooms. This is not because the experience of a spectator in a living room will be free from such technology; rather, it is because the cutting edge of delivering new experiences through digital may only be offered within highly specialized venues. Again, I am mindful of the shifting places in which technology has relocated physical activity—from Straub's treadmill to Nintendo's Wii. The spaces we occupy to enjoy leisure activity change significantly as a result of technological capacities that enable new experiences to emerge. Maintaining the arena experience may also have an important economic role—the importance of ticket sales is one reason for why it will continue to have a function as a communal space.

Yet the complexity of these developments requires further investigation. Recall my mention of the Atlanta Falcons' billion-dollar stadium in chapter 2, where I discussed how "impact seating" would heighten spectators' experiences. It is not difficult to imagine how home-based technology could also create haptic experiences, although the physical proximity of the spectator to the action may be the underlying rationale for providing it within a stadium—spectators can be very far away from the playing field. Nevertheless, a chair similar to the D-Box chairs I mentioned in chapter 2, linked to a digital television, could allow a similar haptic experience—when a touchdown was scored, a home viewer's seat could vibrate to heighten the sensory experience.

The overlaying of technological experiences onto live events is not without precedent. For instance, it is common for live spectators to wear headphones while watching a game so as to listen to radio commentary. Indeed, sports have been transmedia experiences for some time already, where user communities help shape and refine the particular medium platform at any given point in history. Fans decide what kind of technology will enrich their encounter and adopt it accordingly, rather than it being designed by sports producers.

Yet each approach to designing additional experiences has implications for how sports arenas become gateways to non-live audiences, and augmented reality may also complicate certain aspects of this. For instance, the International Olympic Committee's policy of having "clean" venues, with no logos or branding other than the Olympic rings present within the field of play, may be compromised by the increased use of screen-based experiences. After all, if the spectator spends part of his or her time at a sports event watching a mobile device rather than the competition, then controlling what happens within that screen becomes even more crucial for the federation or for holders of television rights. For instance, a third party might develop an app that provides an augmented experience based on what is seen through the camera on the playing field, completely bypassing the rights privileges associated with the sports economies that produce the event. GPS jacking, in which a developer populates a physical GPS location with content unrelated to the owner of that space, is a case in point. In a classic example outside of sport, the artist collective ManifestAR used GPS as a way of invading gallery spaces and turning them over to the public to populate with art work, albeit through an AR (augmented reality) app.

Today sports authorities do little to prevent audiences from utilizing a range of mediating technologies through their mobile devices within venues. However, in June 2016 Apple was granted a patent for camera-blocking technology, which could be used to prohibit filming at music concerts. Yet sports may not go down this route. If anything, sports are increasingly trying to figure out ways to allow audiences to create and experience more within the live context. The question for sports producers is how to do this without jeopardizing the other stakeholder interests that operate around the arena, such as the broadcast rights holder privileges. Yet for the official sponsors and media providers it is a different story altogether; distributing their content via mobile devices is already becoming an important means by which they connect with audiences.

Digital Urban Space and Second Screens

Augmented reality also has a role to play within the staging of sports events beyond just what happens within the arena. This is a crucial opportunity for sports producers to recognize, as a visitor to any sports event may spend only 10 percent of his or her travel time in the venue seeing the sports. Thus, finding ways to connect with audiences during the rest of their time within the event hosting location can dramatically increase the potential for generating revenue and interest. Moreover, what takes place around the sports is mostly underexploited by sports producers, though it should be thought of as a crucial element of the sport spectator's entertainment experience. Recognition of this wider context within which sports events occur is apparent in how the Olympic Games lead city planners to think of their entire region as something of an Olympic theme park during the Games, where they are encouraged to erect all kinds of adjunct experiences, such as the staging of urban screens within Olympic live sites, in which touring sports fans and non-ticket holders can take part in the festival atmosphere.

One of the most predictable forms of digital occupation within the elite sports arena is through sponsorship and advertising. If a sports organization can find a way to monetize some aspect of the live spectator's attention, without compromising the integrity of the experience, then this can increase the revenue around the sport. Indeed, for the Olympic Games alone, the value of the transmission for media broadcasters has grown considerably in the past twenty years, from $1.251 billion in the period 1993–1996 to $3.850 billion in the period 2009–2012 (International Olympic Committee 2014) with a forecast of $5.6 billion for 2013–2016 (IOC 2016).[3] Some of the more typical ways in which this has been achieved is within the competition arena itself.[4] For instance, over the years, the physical billboard within stadiums has gradually been transformed by digital technology, most interestingly through billboard-replacement technology. This innovation allows event producers to tailor the content of a billboard to a specific broadcaster territory. For example, those watching a football match on television in the United Kingdom see different in-stadium billboards than those watching it in South Korea. As Supponor (2015) describes,

Stadium billboards are marked with a special film that absorbs a certain frequency of near infrared light which is invisible to the naked eye. Optics attached to the broadcast camera instantly recognise the billboards and integrate replacement billboards into the live feed. If there are obstructions in front of the billboard, such as players, the system recognises this and masks the "real" figure moving in front with the replacement billboard image behind. It does this in real time, so viewers see an uninterrupted live feed with natural-looking perimeter billboards. Fans inside the stadium see the original billboards which appear completely normal.

The digital billboard replacement pioneered by such companies as Supponor offers even more opportunities to create targeted sponsorship packages for global audiences. Yet it also introduces a further dimension of unreality into the stadium—where billboards have become screens within screens, televisions within the television (Love 2011).

The external spaces of sports venues have also become media zones in their own right, thanks to digital technology. For example, in recent years the rise of urban LiveSites mentioned above, as places for audiences to converge and watch sports events, have created additional spaces in which people can enjoy closer proximity to live sports events. A notable example of this is an area outside the Wimbledon tennis venue where spectators can sit and watch a large screen broadcast of what is happening within the tennis courts during the Grand Slam. The area became prominent when the British player Tim Henman was enjoying success on the world circuit. It now functions as an additional, ticketed venue, the focal point of which is a large screen. Other large-scale events have also given rise to the creation of urban screens, which function as additional venues for people to gather and into which additional services are provided. At the Olympic Games these so-called LiveSites have been prominent since the 2000 Sydney Games— broadly speaking, since the technology has been available. Beyond simply delivering content to new audiences, these new spaces allow organizers an effective way of managing crowds in cases where there is expected to be high attendance. This is particularly important in the production of major events at which there is a need to manage large numbers of people.

In some cases, spectators who do not have tickets for events may come to a city simply with the hope that they will gain access to a LiveSite and take part in the mega-event city celebration. Indeed, often the wait to enter one of these auxiliary spaces can take hours (Piccini 2013). These digitally mediated public spaces have become competition venues in their own right. Around the live sites, additional opportunities for event stakeholders are provided in the form of showcasing, or providing entertainment opportunities and the investments into these temporary buildings can be extensive.

Effectively, they are an additional way of extending the brand of an event, albeit under slightly looser regulatory conditions. These venues go some way to enabling wider participation in the mega-event spectacle, but their history is intertwined with the sponsor pavilions at the Olympic Games, which aim to engage the public around a series of brands, which have made an economic investment in the Games and which are seeking to recover part of that through Games time exposure. At the 2004 Athens Games, the 2008 Beijing Games, and the 2012 London Games, LiveSites were also set up in *other* cities within the nation to extend audience engagement. For the 2004 Athens Games, the city of Thessaloniki was decorated with the host city's branding assets—banners, and so on—and LiveSites showed sports, as also was done in Shanghai for the 2008 Beijing Games and in many cities around the United Kingdom during the 2012 London Games. In a digital age, the capacity to provide additional marketing/engagement opportunities is extended considerably through these spaces. The ironically named LiveSite (ironic because the central event is mediated, not live) is a consequence of the digitally mediated age, though its future may be yet compromised by the growing capacity to deliver the live content straight to mobile devices, rendering the physical space in need of some re-configuration. If the LiveSite at a mega-event is to have a future, it may need to have its own live content to attract audiences, relegating the principal intellectual property of the event—the sports competitions—to a secondary position.

At the 2012 London Games, an innovation in digital advertising took place within small-scale billboard technology, notably the small posters that sit alongside the escalators that go down into the London underground trains. Historically these posters were printed physical matter, but in advance of London 2012 some of them were replaced with digital screens, making the content far more dynamic. One of the most interesting aspects of this was the nearly real-time update of the screens with Olympics-related information. For example, one of the national Games sponsors was British Petroleum (BP). On its underground billboard screens, within advertisements, it would provide updates on the latest Team GB medalist the morning after a victory. Again, such interventions make it apparent how additional ways of monetizing Olympic program content is made possible by these enriched forms of digital advertising, transforming the incentive to commit advertising spend to locations that, were it not for the technology, would remain a relatively flat form of expected revenue generation. Furthermore, it also evidences the manner in which the Olympic Games can be a catalyst for innovation that can benefit a city after the Games have concluded, albeit in the form of generating revenue.

Digital innovation compels us to think about which elements of any given sport are crucial to the sport spectator experience and which aspects could benefit from digital augmentation. This does not require making every aspect of the spectator experience digitally dependent; instead it requires understanding how the spectator experience can be enriched by a digital solution. In part, this chapter's inquiry is connected to a longer history of creating immersive experiences through mediation or performance—from theater to augmented reality. The history of mediated sport reveals how changes to the spectator experience are not equally well received or even considered to be enhancements of that experience. Furthermore, once spectator experiences become encoded, implementing a transformation to that carefully balanced experience may jeopardize their enjoyment, as the example of the FoxTrax puck suggests. Nevertheless, immersive technologies such as Oculus Rift and Google Glass reveal how the relationship between the athlete and the spectator can become even closer.

Also at the heart of the digital transformation of spectator experiences is a debate about the value of the live experience—the importance of being an eyewitness to an event. What it means to be somewhere is also disrupted by the existence of virtual worlds. Although one might imagine that tele-visually mediated experiences fragment and distort the realness of a sports performance, divorcing the spectator from the game's natural spectatorial form, this raises the question as to what a real, integrated, natural, or true version of the game is, and how we would know it to be so. As Turner describes (2013, p. 89), in a world of mediated live events "the mediated 'live' almost becomes a 'third order of simulacra' replacing the original real 'live' event itself, simulating an almost 'realer than real' world in which the event is situated." This chapter has sought to challenge such assumptions and has examined how mediated and even live first-person forms of spectating are all creative, synthetic human constructions, no single version being able to claim greater proximity to the true experience, or the way sport really is, than any another. It has examined those digital technologies that are creating interactive television programming, making it possible for spectators to adopt the role of television producer, selecting camera angles, as well as replaying formats and speeds, to suit their interests and needs.

In the future, the sports industry needs to re-imagine the sports spectator's journey through their events, considering the different spaces in which a spectator may be engaged by the producers of their principal experience. From the moment someone buys a ticket for an event—and perhaps even before that—he or she embarks on a journey that can be enriched by digital content. Whether they are sitting in their home and seeing advertising on their television, or on their way to the event itself, these are all moments rich with the potential to engage an audience. Finding a way to do this effectively through digital technology may open up new opportunities to add value to that journey while also providing additional revenue streams for the sports community.

III The Olympic Games and Sport's Digital Revolution

To explore the application of media change in context, the chapters in this part examine in detail the case of the Olympic Games, which have always pioneered new media technology. Recent examples reveal interesting insights into what is coming next for Sport 2.0. In October 2009, the International Olympic Committee convened its 13th Olympic Congress, a showcase of current Olympic priorities and future direction. It took place within the IOC's 121st session, at which the 2016 Olympic Summer Games were awarded to Rio de Janeiro, making Brazil the first South American nation to host the Games. One of the Congress' main themes was "The Digital Revolution," and the discussions brought into focus the IOC's long historical—and financial—interest in media technology. It also generated debate about some of the issues that arise from developing such relationships. Two central issues were highlighted, and their resolution has paved the way for the next era of the Olympic movement's relationship with media technology and culture.

The Olympic Games provide a particularly interesting case for studying digital change within the elite sports industries. Indeed, over the years, the IOC has been at the forefront of media negotiations and rights ownership, and the Olympic Games have always been a showcase for media innovation, and a community of elite media organizations in which they can experiment with new technology. During its 2009 Congress, the IOC asked two difficult questions about how the Olympic Movement would nurture its rich media culture. The first question concerned how the IOC would come to terms with seeing itself as a *media* organization, rather than as just an organization that relied on selling contracts for the exclusive use by media organizations. The second concerned the means by which this would be realized, which will require much greater engagement with current digital trends—notably social media, mobile broadcasting, and (broadly) Web presence. The IOC's final report advanced a number of conclusions, which are helpful to outline in detail:

5: The Digital Revolution

... The Olympic Movement and its members must be fully cognisant of the impact of this development on all its activities. Future strategies and approaches must be planned in accordance with the massive new opportunities and changes brought about by the digital revolution.

59. A new strategy should be defined to enable the Olympic Movement to communicate more efficiently with its own membership and stakeholders as well as to allow for effective information dissemination, content diffusion and interactivity with the global population, in particular with the youth of the world. It should be an integrated strategy which includes the full coverage by all media and in all territories, of the Olympic Games, as well as the recognition of the new opportunities to communicate the fundamental principles and values of Olympism through all media.

60. The Olympic Movement must position itself to take full advantage of all opportunities offered by the digital revolution, information technology and new media so that the fundamental inherent values and objectives of the Olympic movement are reflected, while the rights of the IOC and the promotion of the Olympic Movement are protected.

61. In order to disseminate the values and vision of Olympism, the IOC and other stakeholders of the Olympic Movement should undertake a fundamental review of their communication strategies, taking into account the fast-moving developments in information technology and, more recently, the digital revolution.

62. The IOC and all other constituents of the Olympic Movement should explore all possibilities offered by the digital revolution; ensuring the broadest coverage of the Olympic Games, including the Youth Olympic Games, as well as of all other games and other major international sport competitions recognized by the IOC or to which the IOC has granted its patronage.

63. The IOC and all constituents of the Olympic Movement should give special attention to the opportunity provided by new technologies to gain increased penetration, exposure and greater accessibility worldwide.

19 / 20

64. The establishment of a Digital Task Force including the IOC and other stakeholders is recommended; with a mandate to optimise the development and exploitation of digital technology.

65. The IOC and constituents of the Olympic Movement must recognize that despite the emergence of a new digital age, the widely varying rates of adoption of these technologies are at a different pace in different regions and among different populations. As part of its obligation to ensure the widest possible global reach, it is therefore important that this is addressed and that appropriate technologies are used to ensure that all have access to the Olympic Games and Olympism in a legitimate and equitable manner and that the issues presented by the digital divide are addressed.

66. The Olympic Movement should strengthen its partnership with the computer game industry in order to explore opportunities to encourage physical activity and the practice and understanding of sport among the diverse population of computer game users.

These conclusions arose in a period during which the IOC had already taken steps toward its immersion within this new digital world. For instance, in the same year, it appointed its first Head of Social Media, Alex Huot, who remains in this role at the present time. Since then, the IOC has launched the development of an Olympic Channel, which functions like a television channel but is mobile first and which delivers content about sports and the Olympics during the time between Games, as a matter of priority.

Nevertheless, perhaps the first question that arose from the IOC's insights was this: If one accepts that the IOC has always been at the cutting edge of technology, then is it doing all it should do to ensure it maintains this position? There are two responses to this question, one is a superficial and one deep, but each is compelling. The superficial view draws attention to the manner in which the Olympics give rise to vast volumes of traffic on social media, an indication of the Olympic industry's central position in the world of sports. The deeper view looks at a number of indicators— economic, cultural, political, and technological—that reveal how close the IOC is to the cutting edge of media change. The answer to this question may be discovered by examining how the IOC has positioned itself around media organizations and how it considers its role in the realm of direct public participation as a media producer. Indeed, these conclusions do not indicate how the IOC considers the Internet, as a means through which the Olympic values can be conveyed to more people. After all, the conclusions emphasize the desire to "communicate more effectively" and to "disseminate the values of Olympism."

Throughout part III, I will broaden these aspirations and explore the Olympic industry's relationship to digital change while also inviting questions about who should be recognized as the main producers of media content in a digital, social world. I will appeal to the idea that the IOC should look at the Internet as not only a means of communication but also a space through which its values can become critically evaluated and more meaningfully owned by the expanded Olympic family, including sports fans, amateur athletes, and people who are committed to the idea of sports' changing the world for the better. In short, I will outline the idea of an *open-source Olympics*—an idea that articulates a third era for the Olympic movement and aims to embody the values of the Olympic Charter more effectively.

I begin by inquiring into the role of the Olympic media in establishing the Olympic movement as an affluent powerful force, partly through its relationship with the media. Next I consider how the Olympic media have changed in the digital era, both in terms of the culture of media production and journalism and in terms of the population and the methods of reporters who cover the Games. Next I examine the opening up of the IOC to contracts with new-media organizations, such as Google, Twitter, and Facebook. This encompasses a debate about how sports organizations can monetize user-generated content to create revenue for the IOC without jeopardizing its current financial base. More broadly, it considers how the rise of social media should provoke a shift in the administration of sport. Finally, I focus on how citizen journalism and guerilla-style media practices have transformed traditional definitions of ambush marketing. Overall, part III attends to the culture of media production that surrounds elite sports, focusing on the Olympic Games as an entry point to discussions about the importance of recent innovations and trends in how sports are adjusting to the digital era. In many respects, insights from this recent history of digital innovation at the Olympics provide the foundations for a world in which virtual-reality sports can become a reality.

6 Media Change at the Olympic Games

In the past hundred years, the modern Olympic Games have stimulated technological innovation within media production. The intimate relationship between the Olympics and the media reveals how the Olympic values of excellence in athletic performance pervade the administrative culture of the world's most watched event. The Olympic motto "citius, altius, fortius" has become a motif for the way in which the Olympic program tends toward transcendence. Every two years, the Olympic Games function as a showcase for their global sponsors and a test bed for their media partners' experiments with the latest technology. This makes the Games an appealing vehicle through which companies can develop their brands with the unofficial Olympic value of innovation.

Through television, slow-motion replay, underwater film capture, mobile tracking devices, and other innovations, media technology has been tested in the context of the Olympics for many years. Three-dimensional and ultra-high definition television broadcasting were showcased at the 2012 London Games, drone cameras were featured at the 2014 Sochi Games and VR/360 film at the 2016 Rio Games. By studying these developments, one learns about the intimate relationship between sports organizations and their consumption via the media. Furthermore, their intertwined histories reveal how innovation is inextricable from the cultural shift toward digital mediation, which occurs alongside technological change.

Numerous events in modern Olympic history tell a story of media change. For example, scanners have been used to develop swimsuits designed to enable the wearer to swim faster and running shoes designed for an athlete's unique foot shape. Such innovations also bring new aesthetic qualities to sports as athletes' performances become increasingly shaped by the technology they use in competition. Equally, media change has altered how the Olympic movement undertakes its work and how it is financed. For instance, the 1980s saw a transformation in the IOC's relationship with broadcast media which ensured that the sale of broadcast rights would also benefit the IOC rather than just the host city (Preuss 2006).

Today television broadcasting remains the dominant medium of the Olympic Games in both financial and cultural terms. Revenue generated from the sale of broadcast rights constitutes 74 percent of the IOC's income (IOC 2016). To the extent that the Games are consumed principally by television audiences, scholars of media and sports have mostly focused on how television reports the Games. Though there are no precise figures on how many people view the Games, credible opinions range from around 2.4 billion to 4.7 billion. Official viewing figures from the 2008 Beijing Games indicate that 61,700 hours of television were broadcast and were seen by 4.3 billion people, reaching 220 territories (63 percent of the world's population) (Sponsorship Intelligence 2008). The same source indicates that the most widely watched event of the Beijing Games was the opening ceremony, which reached 1.5 billion viewers. The 2012 London Games generated 99,972 hours of television and reached a global audience of 6.8 billion, and it was projected that 3.6 billion viewers saw at least a minute of television coverage (Sponsorship Intelligence 2012). However, the locations in which audiences consume the moving image are changing.

Since the year 2000, various factors have ensured that the Olympics are experienced online as well as on television. Since Beijing 2008, the sale of television and Internet broadcast rights has been organized under separate contracts. In principle this means that a national holder of television rights may not be the same organization as the holder of Internet rights. However, in the few cases in which this system has been in place, they have mostly been the same, as often the host broadcaster is also the organization best equipped to deliver a compelling online broadcast package. Indeed, the IOC is moving closer to a situation where television and digital rights are packaged together for a single rights holder. However, it is too early to tell whether this is the right direction for the Olympic movement. Separating rights may give rise to a new generation of online broadcasters, which operate in parallel with or instead of television broadcasters. An indication of this is Amazon's recent acquisition of the gaming broadcast platform Twitch.tv, which has quickly become the preferred broadcaster for e-sports games, attracting live audiences of more than 8 million for single games. Even though the consequences of this shift are uncertain, changes are afoot as a result of the desire to broadcast exclusively online. For example, at the 2008 Beijing Games the IOC set up contracts to deliver content to 78 territories via YouTube, for which there was no other broadcaster (Xiong 2008), while in Europe the main Internet broadcaster was Eurovisionsports.tv, covering a further 72 territories. NBC remained the rights holder for the United States and deserves special mention for its significant contribution to the IOC's revenues. For South America, Terra was the major provider, with 20 territories. TV New Zealand covered 14 territories (ibid.).

One of the questions facing the future of the Olympic media is whether the convergence of technological formats will lead to a collapse of the isolated televisual experience. Indeed, there is already evidence of this occurring, as smart televisions deliver a range of app-based experiences beyond the televisual channels, streaming content through broadband connections rather than television aerials. In the digital era, there is no meaningful distinction between Internet-ready media centers and stand-alone televisions, the latter of which can access a digital television signal or a broadband connection.

Television technology has already changed beyond recognition. They are no longer large, deep boxes with cathode-ray tubes inside, they now are more like computer monitors, thin, sleek, and digitally rendered. In the future, screens maybe simply dumb digital displays absent of any operating system, reliant simply on mirroring content pushed from some other mobile device. Although the content-delivery infrastructure matters more than the apparatus when one is assessing whether change is occurring, it is difficult to ignore how new habits emerge around new forms of content consumption, which can have a cumulative affect on a broadcaster's audience share, if it fails to adjust to these changes. This may be especially apparent within the conversations around the IOC Olympic Channel, launched on August 21, 2016 at the close of the Rio Games. However, the shift toward mobile media may be the biggest change now occurring in how people engage with screen-based content. Moreover, an era of flexible screens wrapped around every imaginable surface is upon us, and this will ensure that media content is pushed out to every imaginable location and is tailored to each specific context and the people nearby.

This chapter explains the context for media change within the Olympic industry as a way of developing an understanding of how technological and cultural innovation have influenced the Games and how, in turn, the Games have transformed societies. It explains how journalists operate at the Games, the media structures that surround the Games, the rise of new media, how the Games may change the way that the media operate within a nation, and the pervasiveness of creative media products that are now constitutive of the delivery of the Olympic Games. And it reveals why looking after the media at the Games may be even important than looking after the athletes.

Situating the Olympic Media

The Olympic media fall into two categories, news and entertainment, and the media contracts issued by the IOC shape the conditions in which sports journalism takes place at the Games, particularly for rights-holding television broadcasters. This does not mean that the contracts necessarily undermine editorial freedom or even journalist integrity, but in terms of the process by which stories are found and explained by journalists at the Games, the conditions of the Olympic media operations determine a large part of what kind of media coverage takes place. Journalists arrive in the host city pre-accredited by the IOC or the organizing committee and then find themselves working under extremely time pressure circumstances. They rely on being fed stories by the Olympic staging machinery of the organizing committee, most of which derives from the sports events.

This system was created by the media; the IOC didn't impose it. This is important to understand when attempting to locate any critique in the appropriate place. Moreover, during each Games, the rights-holding media work with the IOC to improve their conditions for subsequent Games and the pressures on journalists have continued to grow, as have the demands made by them on the Olympic industry. Indeed, over the years sports federations have sometimes expressed concern that the media's agenda has dictated what happens in the Olympic program all too much. Yet today's media are under greater pressure than every before. The stakes are higher, and there is more history to capture than in previous times, but with reduced staffing. The proliferation of new digital platforms means that reporters have to serve as producers, presenters, writers, and editors in the course of their work. Television or print reporters may spend nearly all of their time in the Olympic city within a media center, or at a specific event venue, rarely roaming out into the host city looking for stories. Indeed, the idea that professional journalists should be able to undertake investigative work during the Games is nearly nonexistent and wholly reliant on whether an outlet has sufficient resources to cover these additional stories.

Successful management of the Olympic media is crucial to the success of the Games, but there may be different interpretations of what success looks like. In one version of the circumstances, the Olympic media infrastructure inherently restricts a journalist's freedom to report, since their every movement is orchestrated from the moment they land to ensure that they fulfill a specific expectation to focus on reporting in a way that elevates the Olympic values, made apparent largely through the sports competitions. Moreover, the importance attributed to this aspect of the Games can be seen within all departments of the organizing committee.

From the IOC's perspective and from the perspectives of the organizations that send reporters to cover the Games, everything else is subordinate to reporting the sports competitions—and any other news is considered an impediment to the core business of reporting the sports events. On this interpretation, there is no alarming conflict of interests among the media that are present at the Games. It is not that journalists are undermining their integrity when covering the Games. Indeed, those outlets that are able to dedicate human resources to cover the Games are not obligated to cover anything other than the sports, since this is the focus of assignments.

However, media organizations have more to do than report on sports competitions. After all, the Olympic Games give rise to a number of political, cultural, social, and economic discussions, all of which engage the world's media and are in the public interest. In that sense, a sports media outlet is never simply a sports news outlet. Yet a news media outlet that fails to dedicate resources to other aspects of the Olympic program may be failing to live up to its role as a news organization. The ethical obligation to report what is in the broader public interest becomes secondary to the tacit agreement between the media and the IOC—or that is how it would seem on the basis of the content that is typically created around the Games. For this reason, there may be a need for the creation of an Olympic media watchdog that would assess the freedom of reporters at the Games by evaluating the range of outputs that their organizations generate.

However, there are reasons to be wary of concluding that the editorial freedom of the media is undermined by the Olympic media machine. First, such freedom is protected within the distinction between news and entertainment media. Within any large media organization there will elements of both news and entertainment, but in legal and financial terms their character, entitlements, and obligations are quite different. Indeed, the rules that affect the use of Olympic intellectual property are different for the news elements than for the entertainment elements. News journalists have certain freedoms to report that entertainment media do not have. If a non-rights-holding television broadcaster wants to use some Olympic footage in its evening news—that is, to cover the Games as a news item—it can assert editorial freedom, and the IOC then provides it with a limited amount of content for the purpose of news reporting. However, if a broadcast outlet wants to use content from the Games within a slot that is reserved for entertainment, then the rules that govern non-rights holders will come into force.

Nevertheless, there is reason to be cautious about concluding that all is as it should be in Olympic news reporting, in part because there is a disciplining effect upon the Olympic media, which the Games and the organizers enable and manage. So, while news and entertainment may be separately managed by any media outlet, the influence each has on the other is a crucial consideration. Indeed, the evidence of this disciplinary bias toward self-censorship is apparent when unexpected incidents happen during the Games (as they did at Munich in 1972 and at Atlanta in 1996). In such instances, the sharp contrast of the Olympic coverage with the incidents reveal the narrowness of the Olympic reporting lens. This was also apparent at the 2010 Vancouver Games, during which a Georgian luge athlete lost his life during a practice run. These moments of news reporting are, of course, significant in themselves, but they are even more so because of the narrow lens through which the media frames what is appropriate news during the Games. This contrast is such a break in expectation for audiences that they become moments of profound sadness and emotional outpouring. The same is true of important moments in Olympic history, such as when athletes from North and South Korea entered the Olympic stadium at the Sydney 2000 Opening Ceremony, with their two flag bearers holding the pole of a single Korean "unification flag."

This second view of the Olympic media is more generous toward both the media organizations and the IOC. It does not presume that the Olympic industry seeks to control the media, even if it sometimes endeavors to close down certain lines of storytelling. Indeed, this happens in quite pragmatic ways. For example, on the February 20, 2014, during the 2014 Sochi Games, the IOC's daily press briefing involved one reporter asking the IOC's Communications Director whether the Ukrainian team had asked permission to wear black armbands during competition to acknowledge civilians who had lost their lives in violent clashes in Kiev some days earlier. To understand the significance of this question—and the answer that would follow, it is useful to know a little more about the context. Thus, if an official request had been made by the team or the NOC and the IOC had rejected it, that would have been a major Games-time story. However, the IOC's response was that no official request had been made and the parties concerned were actively trying to find a suitable way to acknowledge the sentiment of the athletes. The implication—not expressed explicitly—was that wearing armbands in the context of a political conflict might have the effect of politicizing the otherwise neutral zone of the Olympic competition. In this case, it was clear that the IOC was interested in ensuring the appropriate message would be conveyed. Yet it was also apparent that the IOC knew that it would have been subject to criticism if it had been revealed that an official request had been denied by them. However, in responding to the inquiry, it is not reasonable to claim that the IOC controlled how reporters did their work, or that they sought to suppress any suggestion that an official request had been made. Indeed, the Games-time daily press briefing with the IOC and the Organizing Committee is a key site where this independence is played out.

Nevertheless, the Olympic mechanism gives rise to a system of self-censorship in media agenda setting. The media voluntarily—though as a result of financial and political relationships—locate themselves within a situation in which their capacity to report the Games in full is compromised either because they don't have enough resources to tell other stories or because they have a pre-designed Games-time reporting agenda that focuses only on celebrating the Games while minimizing the use of resources. Crucial to this consideration is the fact that sports events work with captive audiences; it is reasonably cost-effective reporting, since the content is produced by the sports themselves, the schedule is known in advance and news can be easily planned.

While it is likely that each of these interpretations of the Olympic media holds some truth, the relevance of this inquiry becomes apparent when examining what gets reported by the accredited media and, by implication, acknowledging what is neglected. I will consider this later in the book by examining the content generated by citizen journalists at the Olympic Games.

A further distinction that is necessary to make when situating the Olympic media is between different forms of media that create Olympic content. For example, in the case of television, the Olympic Games leads to the creation of new commissions the purpose of which is to build audience engagement with the Games ahead of them taking place, which, in turn, aims to increase Games-time viewing figures. Thus, before the Games it is common for television channels to create content that draws on public interest in the social significance of the event, elevating its political and cultural importance. Admittedly, it is difficult to separate out what may be justified on the basis of public interest and what is designed to build loyalty to an Olympic broadcaster, since they are often one and the same. For instance, this programming sometimes takes the form of documentaries about an athlete's journey toward Olympic competition. Alternatively, it can involve quite sophisticated pieces of investigative journalism. For example, during 2007 the UK's Channel 4 produced a number of programs about China in which the Olympic Games were used a backdrop. Similarly, the BBC's flagship investigative journalism program *Panorama* has produced several documentaries about Olympic corruption, notably around the period of the bid for the 2012 London Games, where they endeavored to expose corruption in the Olympic city bid process. These examples of how Olympic programming enters the public domain via the media do not always directly involve the IOC or even the Organizing Committee, but they cohere with the broader programmatic goals of the broadcaster, or the editorial priorities of a newspaper. Yet collectively all of these stories become part of the ongoing Olympic narrative, which is repeated every two years via the landmark of the Games themselves.

While drama and documentary tend to be the formats through which broadcasters build awareness and interest in the Olympic journey, comedy has also featured as a vehicle for driving interest in the Games. For example, the mockumentary television series *The Games*, produced by the ABC network in 1998, was particularly successful and was widely discussed as having broken new ground in Olympic programming history—and in comedy history too. It told the story of a fictional organizing committee for the 2000 Sydney Games in a satirical manner. It was so successful in Australia that its stars—as characters from the show—ended up participating in the Sydney Games' closing ceremony. A decade later, a similar comedy produced by the BBC preceded the 2012 London Games. Titled *Twenty Twelve* and mirroring ABC's comedy, it provoked a public clash over the intellectual property of the idea, since the ABC creative team wasn't involved with *Twenty Twelve* in any way. In each case, there may be something particularly interesting about the role of comedy in generating public support for politically controversial initiatives, in some cultures (Plunkett 2011).

Although it is tempting to talk simply about the comedic merit of *The Games* and *Twenty Twelve*, their satirical style invites deeper analysis particularly since the Olympic Games are so publicly controversially. Arguably, the political role of satire in these cases serves to dissipate public concern, despite the aspiration of satire to draw attention to issues of serious political concern. For example, in one scene within *The Games*, a shortfall in sponsorship leads the organizing committee to consider approaching a tobacco company as a sponsor. This example refers back to a longer historical context where tobacco has been systematically removed from the sports sponsorship world, because of the negative health associations of smoking. As the story unfolds within the episode, it appears that all the tobacco company wants from the organizing committee in exchange for its money is for the employees to take up smoking and be seen publicly smoking their brand of cigarettes. The message here relates to the willingness of the sports industry to prostitute its values in pursuit of endless advertising revenue, a message that the sports world may still need to consider.

Similar examples are found in the BBC's mockumentary about the 2012 London Games. For instance, one episode deals with the hyperbole that surrounds the use of social-media marketing as an audience engagement device. The organizing committee is planning a marketing campaign that exploits the public interest in the Olympic Games and the British Queen's Jubilee anniversary, which are occurring in the same year. (This is the part of the story that is true.) The creative team come up with the catchy campaign title "Jubilympics," a portmanteau of *Jubilee* and *Olympics*, which was a story line intent on drawing attention to the absurdity and superficiality of social marketing and the impossibility of brand crossover in the tightly controlled world of Olympic intellectual property. The storyline was an attempt to poke fun at the sometimes superficial staging of mega-events and the sometimes pretentious manner in which public interest is gauged and exploited.

In view of this wide range of media artefacts that surround an Olympic Games, defining who is a journalist, what rights journalists have, and how they are served and managed are important in determining which stories get told and who has control of the narrative. This is why it is necessary to scrutinize the notion of the journalist in the context of the Olympic Games and the processes by which citizens can be granted that status. Evidence from recent Games demonstrates how the journalist's role is gradually being transformed by a public who want more from the Olympic media than just information about the sports competitions. Indeed, the concept of the journalist has changed in a digital era and, with it, what is required of the IOC and the host cities. This is the first crucial shift in the era of social media at the Olympic Games.

Media Structures at the Olympics

The official media structures at the Olympic Games are the result of a combination of operational and financial needs. Ever since the Games' financial crisis of the 1970s and the restructuring of the Olympic Movement as a commercially viable enterprise in the 1980s, the IOC has treated the media as a crucial Games stakeholder and a key member of the "Olympic Family," which includes international sport federations, the athletes, team officials, sponsors, and IOC guests. To secure full coverage of the extremely diverse and concentrated range of Olympic activity during the sixteen days of competition, the host city is required to provide members of the media with state-of-the-art working venues,[1] a fully equipped Media Village providing meals and accommodation, transportation to all official Olympic venues coordinated with the times of competition, and an extensive network of information points with the latest updates on all sports events and competitor backgrounds.

To control the number of media organizations with access to such facilities, the IOC has set a strict accreditation process similar to that established for the rest of the Olympic Family (International Olympic Committee 2015a, Rule 52). For press writers and photographers—the professional sports spectators—the IOC has set a maximum of 5,800 places per Games since 2000; numbers are allocated per country, with priority given to the main media organizations, which are determined by respective National Olympic Committees. Broadcasting organizations, as the main funders of the Olympic Movement,[2] are treated differently, since television is the medium that has allowed the Olympic movement's finances to flourish. Thus, broadcasters are not only treated as accredited media, but also as Olympic rights holders with access to the core Olympic properties, such as the rings, for use in marketing materials. The IOC states that "rights are only sold to broadcasters who can guarantee the broadest coverage throughout their respective countries free of charge" (ibid.) and they are offered exclusively to one broadcaster per geographical area. This means that in any one country there is typically only one approved official broadcaster and no competing TV channels can offer moving images of official Olympic events, except for editorial purposes. Broadcast organizations are allocated a set number of accreditations according to the level of funding support. Today the total number is approximately 14,400 individual accreditations, including reporters, producers, and members of the technical staff.

There are two main media facilities at the Games—the Main Press Centre and the International Broadcasting Centre, which operate in slightly different ways. Access to the latter requires a higher level of security clearance, as it holds the strictly protected "moving-image" feed of all sport competitions, which are available exclusively to right holders. Nevertheless, both of these venues are typically connected physically and share two characteristics as the main official accredited media venues. The first is that access to each requires full accreditation (or guest passes, which can be organized by an accredited person) under strictly limited quotas. The second is that they focus on providing information only about official Olympic events, including the Torch Relay, the opening ceremony, the official sporting competitions taking place during the sixteen days of the Games, and the closing ceremony.

Holders of Olympic broadcast rights often have access to all Main Press Center facilities, while the press and photographic media cannot enter the International Broadcast Center. Non-rights-holding broadcasters may be entitled to apply for accreditation at the Main Press Center to access and distribute text-based information about official events, but, as in the case of the press, they cannot gain access to the International Broadcast Centre or any moving images. This stipulation also encompasses the distribution of such images in an online environment.

The Olympic Charter specifies the IOC's commitment to protecting the media coverage of the Games, as well as outlining the technical regulations imposed on journalists for this purpose (International Olympic Committee 2015a, Rule 48). In particular, it identifies the objective of the IOC to maximize media coverage and for such coverage to "promote the principles and values of Olympism" (ibid., bye-law 1). In so doing, the IOC asserts its authority over the media's governance at each Games, by orchestrating daily press briefings, for example. Moreover, the host city is bound by these requirements, as an integral part of its Host City Contract. By extension, the IOC also asserts its exclusive rights by stipulating that

Only those persons accredited as media may act as journalists, reporters or in any other media capacity. ... Under no circumstances, throughout the duration of the Olympic Games, may any athlete, coach, official, press attaché or any other accredited participant act as a journalist or in any other media capacity. (International Olympic Committee 2015a, Article 48, By-Law 3)

Accompanying these provisions is a series of other measures that aim to control the circulation of images to Olympic audiences. Host governments, at the behest of the IOC, are compelled to institute novel legislation that will govern the protection of the Olympic properties. These unprecedented conditions often attract criticism for creating a situation in which prior human rights may be undermined. Indeed, the Canadian filmmaker Jason O'Hara created a citizen-led documentary about the 2016 Rio de Janeiro Games, titled *State of Exception*, to describe precisely the range of consequences that arise by governments instituting such policy around the Games.

"State of exception" is the phrase used to describe what happens when the government changes the law in order to deliver on the nation's commitment to delivering the Games and it is not without controversy or consequence. For example, for the 2012 London Games, the British Government instituted an "Olympic Bill" (House of Commons 2005). These stipulations reveal that the IOC asserts the Games as its legal intellectual property, shared temporarily with the host city and its stakeholders. This indicates a division in the direction of the Olympic narrative—between the IOC, which is setting the conditions of the stage, and the host city, which is facilitating its orchestration. These conditions have led to precise stipulations in the contracts, which protect against the infringement of that property. For example, Olympic Charter Rule 53 states the following:

... 2 No form of advertising or other publicity shall be allowed in and above the stadia, venues and other competition areas which are considered as part of the Olympic sites. ... 3 No kind of demonstration or political, religious or racial propaganda is permitted in any Olympic sites, venues or other areas (International Olympic Committee 2015a, Olympic Charter, Rule 50, p. 93)

The effects of such guidelines also apply to persons who enter Olympic venues, including spectators, athletes, and officials. They are prohibited from doing or wearing anything that might act contrary to this Rule. These areas of IOC regulation continue to expand from one Games to the next. For instance, at recent Olympic Games, all billboard space within the city center and areas surrounding the Olympic venues has been offered to Olympic sponsors or else left empty to avoid "ambush marketing." The consequences of this rule were most apparent at the 2004 Athens Games, during which many billboards around the city were left completely empty. In this sense, the entire city may be construed as an Olympic venue. Here we have, not the Disneyfication of everything, but the Olympification of the entire public sphere, a complete occupation, which itself invites resistance from disenfranchised local communities. Indeed, it is difficult to walk around an Olympic city during the Games and not have some Olympic sponsor message always within one's line of sight.

These conditions are indicative of the manner in which Olympic media content is controlled across the entire Olympic program. From managing reporters to determining which images are shown around a city, there is a remarkable amount of control that takes place around the circulation of media content at the Olympic Games. But although this version of the story focuses on how the media are under the control of the Olympic industry, it is also important to recognize how the influence exerted by the Olympic industry can bring about media change for the better, even when this is hard to verify. Although media change is an integral part of Olympic history in technological and cultural terms, it also has political consequences.

How the Olympic Games Change the Media: The Case of China

Although debates about the Olympic media often focus on how they have changed our experience of the Games, the reverse is also sometimes true. The staging of an Olympic Games can bring about changes in how a nation's media operates, or how the nation works with foreign journalists. It may also have a significant impact on how a broadcaster operates. For example, in advance of the 2012 London Games the BBC convened an extraordinary steering group consisting of heads of programs from across the organization. The role of this group was to ensure that various departments and key programs within the BBC would have a joint approach to creating Olympic content in the years leading up to the Games. The BBC also instituted a program of journalism apprenticeships focused around the Games, which was also a new initiative for the organization.

Sometimes the Olympic media's impact on society can be felt at a relatively low level—for instance, in the way that the Olympic program may lead to new forms of collaboration within the media industries, as the OBS demonstrates. The OBS is a fascinating organization, worthy of further study to understand how the Olympic program creates a labor community whose work stretches well beyond simply covering the Games.

One distinct element of the Olympic media machine is that it creates a situation in which an unprecedented volume of journalists come to a city, most of whom are overseas reporters. Each of these reporters arrives with his or her own sense of journalistic ethics and working ethos, and yet they are all expected to fit in with the particular rules that govern journalism within the host nation during the Games. In some cases, the way in which the journalists work in an Olympic host nation may be significantly more liberal for visiting journalists than what they experience in their home country, or it may be considerably more restrictive. In each case, the coming together of journalists as the Games—along with the liaison that takes place on the approach—is an opportunity for promoting intercultural dialogue about the role of media in society. The potential of this network may yet to be realized fully, but sports events are among the few regular planned media events for which such continuity of organizing is possible and for which strategic thought about the role of the media in social change could be a core element of what the network does.

The Games is also a platform for analyzing where the media and politics interface and there have been some crucial Games where tension between these two sectors has brought with it interesting transformations in how a nation relates to the rest of the world. The most recent, prominent example of this is the Summer Games of Beijing 2008, where the political debate about media freedom and how the growth of new media in China would be managed was, in some sense, transformative. The coincidence of China's first Games occurring at the time of a social-media revolution, set the stage for an enormous range of possible changes to China's media policy. It also shined a light on global concerns about China's governance over the media generally and the Internet specifically. This period also encompassed discussions about censorship surrounding the presence of Google in China, a debate that would return in the post-Olympic period in the context of Twitter and resurface again around the Nanjing 2015 Youth Olympic Games. Whereas before the Games, China (particularly Beijing) espoused open media policies for journalists, after the Games were over discussions about surveillance and restrictions again emerged.

Yet China's online population was beginning to boom. The statistics on mobile and online access were staggering. In 2007 there were more than 480 million mobile phone users in China, 17 million of whom use their phones to access the Web. The number of Internet users in China reached 137 million, exceeding the number of users in the United States by February 2008. In Beijing alone, there were nearly 5 million Internet users—30.4 percent of the city's population. The percentage of Internet users in the under-30 age group reached 72.1, and the number of bloggers in China reached 20.8 million (China Internet Network Information Center 2007; Weitao 2007). Thus, a large proportion of the population were able to access the Games in ways that differed from those offered by traditional broadcasting.

It is no easy task to locate the origin of the Olympic influence on China's media policy, or to identify its end point, but one landmark was March 2007, when the IOC launched a tender for the sale of the Internet and mobile platform exhibition rights (new-media rights) to the 2008 Beijing Games for China's mainland territory. This was the first time that the IOC separated the sale of television transmission rights from Internet and mobile broadcasting. However, while China endeavored to honor its commitment to the IOC by abiding by the rules of Olympic media coverage, some of its own domestic media management laws and regulations had not been upgraded to meet these commitments.

Effective February 2003, China's State Administration of Radio Film and Television (SARFT) instituted the Administrative Measures regarding the Broadcasting of Audiovisual Programs through the Internet and other Information Networks in China, which stipulated that a broadcaster must first apply for a "license to broadcast audiovisual programs by network" before they can broadcast audiovisual programs online. However, many broadcasters' Internet content providers, including Sina.com, Sohu.com, China Unicom, and QQ, did not have such a "license," which meant that they would not be able to broadcast under this regulation. In a transcript from Sohu.com in the first quarter of 2007, the company indicated that it had no role in the delivery of such content:

Sohu is the exclusive Internet content provider sponsor for the Beijing 2008 official website, so we are the operator of Beijing2008.com or .cn, for that matter, and all content on that website is provided by Sohu. ... The new media rights is a separate matter and that is closely tied into with the TV broadcasting rights, so yes, there was a tender but that is separate and distinct from the official website that we operate. So it is almost like TV broadcasting rights in the eyes of the IOC. The outcome of the tender will be known probably—if not during Q2, it will be early Q3. So it is separate and distinct. (Carol Yu, Co-President and Chief Financial Officer of Sohu, cited in Seeking Alpha 2007)

In this moment, China may have found itself in a situation in which a new, digital broadcaster, could enter the stage and become a significant player—alternative to television—in the Internet broadcaster world, perhaps exerting a disruptive influence on the media conditions within China. However, it was not to be. Moreover, for China the SARFT regulations indicated that there would be considerable barriers to a non-China-based company delivering such content. Indeed, it was likely that a number of China-based companies were going to struggle with the regulations. In any case, China-based bloggers would face unknown penalties for broadcasting material via the Internet, though this was likely to be of concern only in the context of moving sports images.

A further consideration that made Beijing's Olympics particularly interesting in terms of media history was the development of digital broadcasting technologies. As Weber (2005) notes, China's government strategy was founded on an interest in maintaining political control while enabling steady economic progress. The challenge for telecommunications had always been that "authorities desperately want to control the flow of news and opinion—especially dissent," while "the government also wants an open, modern and efficient economy, including a state-of-the-art telecommunications and information infrastructure" (ibid., p. 792).

How were these aspirations affected by pervasive reporting in an era in which producer, publisher, and audience had become irreconcilably blurred?

Weber provides one answer indirectly, offering explanations for how different cultural forms are treated. Thus, Weber explains that the media are typically treated as "nation-building" entities in which entertainment enables "consumer support." Perhaps if "new media" publishing in the context of the Olympics is treated as a cultural industry it will not be seen as presenting any destabilizing potential at all, but this would be a naive view to take on the contribution of media artifacts to societal dynamics.

Another moment when the Olympic Games influenced China's media policy occurred on December 1, 2006, when the government instituted a set of regulations granting foreign journalists more freedom to report on the country in the run-up to and during the Beijing Games than had previously been the case. According to China's top publicity official, Cai Wu, "the Olympic Games provide us with a good opportunity to adjust the regulations," though post-Olympics China does not really testify to this having changed very much. Indeed, the regulations came into being on January 1, 2007 and came to an end on October 17, 2008. Similar anxieties about the openness of China's media were voiced again on the lead up to its next Olympic event, the 2014 Nanjing Youth Olympic Games. In that case, even a registered journalist working in the Main Press Centre at the Games might not have enjoyed open access to the Web, which was available only on certain specified intranets and only with certain platforms.

Despite the short life of the legislation surrounding the 2008 Beijing Olympic Games, it is unreasonable to conclude that it had no impact on how China approaches digital connectivity and open communications today. Indeed, the international attention to the debate about domestic media freedom in China, which preceded the Games and which engaged Human Rights Watch and Reporters without Borders in campaigns concerning the treatment of journalists; in addition to the legislation, may have allowed certain reporting of this issue that otherwise would not have occurred.[3] This alone may have influenced the perspective of a generation of people whose views about such freedoms were informed by such conversations. The 2008 Beijing Games was not the first instance of governments drawing up special legislation to cover the reporting of the Olympics, though in the context of China this was all the more important in view of its ascent within the global economic community. These factors mattered most in the context of the non-accredited media population at the Games—that is, those who came to China to cover what happened around the Games but whom may not have been official, professional journalists. These non-accredited media are the new population of reporters at the Games and will be the focus of the next chapter. For now, it is useful to note that this, again, was a moment of disappointment when considering the potential of the Games to bring about media change, since only foreign visitors who had an official journalist visa were given access to the Non-Accredited Media Center, a vast undertaking capable of hosting 11,000 registrants.

The media coverage of the 2008 Olympics could have helped to promote Beijing and China to the world, which is perhaps part of the reason why the Chinese government relaxed press restrictions ahead of and during the 2008 Games. The Games were seen as an opportunity for the Chinese government to implement public diplomacy through new-media coverage via the Olympics Games. Clearly, the Chinese government did not want to lose the chance for the world to know more about its economic successes, especially through the words of people from their own countries. Beijing's hosting the 2008 Olympics offered a great opportunity for China's new-media community. Yet nearly two Summer Games later, one is hard pressed to say that this brought long-term change for the better in China's Internet policies. Nevertheless, a considerable amount of journalism about China—journalism that may not have occurred were it not for the new legislation that was brought about because of the Games—took place over the Olympic period (Smith 2008). Thus, although the impact might have been short-lived, the legacy of the content generated over that period may have been unprecedented, and the long-term importance of that should not be overlooked. For instance, China's media prominence was up by 20 percent in 2008 over 2007 and by 51 percent over 2000, dropping by 82 percent when the Games were over (Anthony Edgar, personal communication). For better or worse, it is clear that the Games offer a unique opportunity to draw attention to a place, especially if there are domestic controversies.

As was mentioned earlier in the chapter, the occurrence of the Nanjing 2014 Youth Olympic Games created a moment to revisit some of these changes. It is a particularly interesting instance to examine, since, despite the absence of global media coverage at these Games, they are still overseen by the IOC, which is an organization that China continues to court. Indeed, the relatively new Youth Olympic Games have already become highly coveted. Arguably, cities rarely take actions that would jeopardize their relationship with the International Olympic Committee, since they

are all always potentially Olympic sites. Indeed, Beijing made a bid successful for the 2022 Winter Games. On the approach to Nanjing, there were
still indications that China would respond to the IOC's request to open up
access to digital communication channels. Indeed, I was involved in the
conversations that surrounded this directly. As a mentor in the IOC's Young
Reporters program, I worked with the IOC's Head of Media Operations,
Anthony Edgar, to convey our Games-time needs, which included access
to platforms we specified during the Games. Although these were not open
across China, and thus the impact on China's media policy is negligible, the
conversation between China's government and the IOC over these matters
is another instance of how the Olympic industry generates conversations.
The consequences of the conversations may be hard to specify, but they can
be seen as a kind of quiet diplomacy. Certainly, until China adopts a more
open media policy, one should be cautious about championing the Olympic Games as a catalyst for change. Yet even the Youth Olympic Games led
to the generation of content of a kind that typically wouldn't emerge from
China, without any editorial restrictions from the IOC or the government.
Such small interventions are, to an extent, historically important.

Olympic Journalism through Social Media

Another way in which media operations have changed at the Olympic
Games is in the use of social media to produce journalism. A crucial feature
of the change is the manner in which media organizations have recognized
the need to take their content to social-media platforms, rather than to
expect to draw audiences away from large social-media environments and
to their own platforms. In the course of a decade of social media, broadcasters and print outlets have had to accept the fact that their own efforts
to create compelling platforms through which they can access the largest
audiences has always been secondary to the volume of potential audiences
found within large social-media sites. Indeed, today, many experiences of
content consumption from third-party providers simply stay within the
news stream of a Facebook user's account, without needing to click exter-
nally to some other website.

In this respect, it is crucial to note the changes that have happened in
the first decade of social media. Initially, social media provided new spaces
for audiences to come together in what had become a fragmented online

world, and old media organizations could reach out through these environments, focusing their investment into online delivery. Yet the second era best describes these platforms as content generators themselves, increasingly monetizing their distribution space through advertising. The old media partners have become subject to a new economic infrastructure. Journalists, having come to rely on social-media platforms in order to reach large audiences, have been compelled to work out how best to make use of the content-generating interests of social media users.

It is now beyond question that, as a distinct social space, social-media platforms are able to generate news that becomes the subject of journalism. The Olympic Games again serve as a case in point. For example, when the "official Olympic protesters" (Malik 2012) also known as "Space Hijackers" created a Twitter account using the London 2012 logo as their avatar, their creative act became a subject of news debate, leading the London Organising Committee of the Olympic and Paralympic Games (LOCOG) to inform Twitter that it considered this an infringement of their intellectual property.[4] This undertaking became a focal point for news attention and is a good example of how social media operate as a vehicle for creating new kinds of performative acts that are newsworthy. I will examine more examples of this later, but here want to briefly discuss how the process of doing journalism has changed as a result of social media. This argument first requires expanding the category of social media to encompass what might be described more broadly as *data-driven journalism*.

The capacity of social media to generate large volumes of data that can provide insights into the world around us has created new skill sets within journalism and new attempts to engage people with content. The rise of infographics as a form of journalistic storytelling is a case in point. Some beautiful examples of Olympics-related infographics based on data generated from various programs have been produced at recent Games. For instance, a *New York Times* animated infographic from 2012, depicting 100 years of 100-meter sprint Olympic finals, displays the winner's position relative to Usain Bolt's world record time and is an effective way of articulating the change within this discipline over the years, in a way that words might have failed to do quite so effectively. Alternatively, the *New York Times'* time-lapse photographs of snowboarders at the 2014 Sochi Games create hybrid visual media, somewhere between image and movie; a veritable digital zoetrope which explains what happens in mid flight during a half-pipe event. Tampering with image through data is indicative of this shifting skill set made possible by data and new design technologies. Further examples of such trends are found in cinemagraphs, which use video material to create still photographic works, which have subtle moving elements. The viewer reads the final work as a still photograph, but elements of the image retain their moving components, to create a hybrid artifact. The work also takes the form of an infinite loop, whereby it is not possible to know when the movement begins or finishes. A range of great sporting examples can be found from one of the industry leaders in cinemagraph software, Flixels (https://flixel.com/cinemagraphs/sport/).The effect is similar to an animated gif in principle, but the combination of using video, freezing a frame and retaining one aspect of it as a moving image, combined with an infinite loop, warrants its being described as a new kind of artifact within our media history (See Lin 2014.)[5]

Today, journalists risk losing their professional edge if they fail to engage with social media and integrate it within their practice, but this is also redefining their contribution. Indeed, at the heart of social media is a pursuit of creative digital alternatives, an attempt from participants within this new economy to distinguish themselves from others—a desire which goes back to early digital years of creating avatars. Today, it is now widely understood that being a reporter—in sports or outside—requires being able to operate across platforms and across skill sets. Toney (2012, pp.118–119) writes:

If you were a journalist covering a leg of the Olympic torch relay in the buildup to the London 2012 Games, you might have needed more than a pen, paper and laptop. At any major event, the demands placed on journalists now have increased. You could be live tweeting off your smartphone and uploading pictures via Twitpic. You might need to be proficient in Audioboo to get short interview clips online or a live streaming website like Bambuser, while YouTube has turned everyone into a broadcaster competing for millions of eyeballs. A laptop and mobile phone with plenty of charge is vital when fi ling on the fly; indeed, back-up batteries are a must because there is no guarantee you will get close to a power source. If you do find a source, it will probably be taken, so make an extension lead part of your toolkit too.

Although there has been some debate as to whether social media make professional reporters unnecessary, evidence indicates that the rise of user-generated content is, instead, leading to a remaking of the reporter's role. For example, Mare (2013, p. 95) outlines how the professional journalist becomes even more crucial to "verification, contextualisation and amplification" in a world of social-media news reporting. Furthermore, Mare notes that the consequence of social-media reporting is the development of a new form of collaborative journalism. Tracking research into the use of social media over time within sports journalism, Reed (2013, p. 558) notes that sportswriters began using Twitter for professional networking and then subsequently for "breaking news, promoting their work, and connecting to readers" and then as a newsgathering service.

On each of these levels, there are challenging implications for sports reporters, particularly since the nature of the live event and the desire for immediate news can undermine the processes by which verification and confirmation of circumstances is possible. Interviews with journalists conducting research at the 2014 Sochi Games revealed the challenges they faced. For instance, there is a need for reporters to verify what has happened before it is reported, whereas a member of the audience may feel no such obligation and may "break" the news item before the journalist has a chance to verify the occurrence. The implications of this are particularly important in unexpected situations. For example, at one event a British speed skater fell and had to retire from the event. At the point of falling, the attending journalists had to make a decision as to whether they tweet the occurrence, or whether they go down to rinkside and find out what had happened and the injury's severity. Doing the latter might risk losing crucial time and not being the first to report the story, but waiting to know more would ensure accuracy of the report. Though there might be some value for the reporters in breaking the news, without details the significance of that news may be undermined. Arguably, sharing only the face value of the incident compromises a journalist's ethical obligation to their audience, since he or she will effectively be telling half of a story—and that half might inspire a certain degree of unjustified panic in the audience. In this simple example cited above, one is alerted to a number of challenges for journalists who are working on reporting live events, where there are also audience members with similar means of communication through social media. Defining the roles of each eyewitness describes the core tension between the citizen and the professional journalist and resolving these differences may require a technological solution. For example, perhaps accredited journalists should be given access to a privileged audio feed that allows them to hear the conversation between the athlete and their team advisers, so they get exclusive, first hand access to what has taken place.[6] Alternatively, perhaps the athlete's body could be augmented with sensor technology that could provide live information about the severity of the injury, so that it is immediately apparent how important the incident is in the athletes' overall career. This could allow the journalist access to the crucial information necessary to provide a near immediate story which extends the capacity of a witness' account. Imagine the reporter sat in their booth and, the moment an injury happens, a screen switches to a close-up, live image feed of that athlete's injured limb, providing medically precise diagnostics on what has happened. In that scenario, whereas the citizen might witness the event and tweet the incident the data available to the journalist would allow him to also say whether the incident is severe or minor. While elements of this proposition seem far removed from what is likely to be available any time soon within elite sports, the growth of wearable technology might soon make this possible. The only question is whether there is any value in maintaining a layer of exclusivity to the accredited media, or whether this should be immediately open to all. In a world where media exclusivity is crucial to the economic foundation of sports, this may be the way to bypass the potential compromises arising from the fact that audiences also have professional media tools available to them to tell stories. Yet access to information that the general public cannot obtain is, in one crucial sense, an organizing principle of journalism, which may explain part of the crisis it faces today, where everything is available to everyone at all times.

The examples cited above also highlight an area discussed by Sherwood and Nicholson (2012) as a tension in how sports journalists use social media—between the individual and the institution. Sherwood and Nicholson note that some journalists "would never break news on Twitter, because it wasn't directly associated with their respective newspaper" (p. 950). Reed (2013, p. 568) also notes that frequency of social-media use may be correlated with professionalism: "Twitter allows them to gather news while keeping a disinterested stance toward the people they cover, while Facebook blurs this traditionally accepted tenet of professional and private boundaries." There is also a sense in which the content of social media becomes part of news syndication directly, with reporters picking up first-person comments within their Twitter or Facebook accounts and republishing them within stories. Alternatively, Artwick (2013) discusses how professional journalism needs to adjust to social media, considering their role as a service rather than as a source of generating end products. Instead of creating stories for audiences, media outlets must think of themselves as tools, through which audiences can create their own stories.

All these transformations suggest that what it means to be an Olympic journalist is quickly changing as a result of new configurations of content syndication and as a result of the rise of a generation of media consumers whose desire for real-time pervasive news content seems insatiable. Though this may not compromise the key components of what journalism entails, it does suggest that additional competencies will have to be acquired in order for that content to remain relevant to an audience with increasingly diverse entry points to journalism content. This is likely to be a feature of the shifting character of journalism over the next decade.

A final consideration in the Olympic industry's media operations is the development of two related organizations. The first is the Olympic Broadcasting Services (OBS), which has, since 2001, been operating as the official Olympic broadcaster, from which rights holders take their televisual content of the sports. The second is the Olympic Channel, a more recent project that emerges from the IOC's Agenda 2020 and is a key consideration within the wider context of Olympic media operations for how it approaches the challenge of media change in present times.

The Olympic Channel launched at the 2016 Rio Olympic Games, aiming especially to deliver video experiences for audiences between the Games, in an attempt to connect with a mobile-first younger generation. The creation of this new IOC platform may be seen as a step toward making sure it is well positioned to deliver a cutting-edge media experience at a time when platforms such as Netflix are transforming the habits of television audiences. Though the IOC has, historically, claimed that is not a content-generating organization, this may be about to change, though it must tread carefully, so as to not jeopardize the interest or good will of the rights-paying broadcasters.

Operated by the OBS, the Olympic Channel introduces new elements to spectatorship, drawing on some of the principles I have outlined around participatory spectatorship and data-driven audiences. For example, it integrates elements of gamification, linking live video content experiences to quiz information. It also integrates third-party applications to make the spectator experience more connected to the trend toward mobile health tracking. For example, a user can share Samsung health tracking data with the mobile app to see how his or her performance in a certain sport compares with that of an elite athlete. Along these lines, it is also possible for the user to select a sound track for his or her training based on the music preferences of their athletic hero. Finally, in a trend toward thinking of televisions as simply dumb screens, the mobile app is able to push video content straight to a television by means of a mirroring function.

Together, these elements of the new Olympic Channel proposition mark it out from what rights holders are doing with their Olympic broadcasts while ensuring that the IOC is placed at the heart of the new-media economy which is driven by the aggregated insights from participant and audience data. This single dimension of the Olympic Channel may be crucial to ensuring that it will flourish in the future. After all, the main aspiration of any media organization today is to become so important to its audiences that their behavior becomes habitually tied to the platform. If the IOC manages to gain the audience's loyalty to its new channel, it may succeed in becoming the Facebook of the sports world, and all broadcasters may then commit even more resources to securing the Games.

7 The New Olympic Media

The prominence of Sport 2.0 as a modus operandi for sports in the 21st century is implied partly by the media ecology that surrounds it. As was noted in chapter 5, elite sports events must be thought of as theatrical performances involving various actors who collectively produce and animate the main event. Such orchestration has always been a feature of sports. Even since the ancient Olympic Games, during which announcers (*keryx*) would speak to the spectators and ritual and ceremony elevated the significance of what happened in the arena, staging has always been crucial to the meaning and the importance we attribute to sports.

In the 20th century, the creative media industries became the central component in this production, but the composition of this community is changing. Mapping out these new media partners in the production of sports events is a crucial aspect of realizing the opportunities arising from Sport 2.0 which are challenging the established model of sports event production.

There are at least two ways to make sense of the concept of new media in the context of the Olympic Games. Moreover, their co-development pertains both to media change generally and to the Olympics specifically as an exemplar of how the global sports industries work with media technology and covet media innovation as a route toward enhancing spectator experiences. First, one may characterize *new media* as the growth of new populations of journalists who have begun to visit the Olympic Games to report what takes place. This population may be described as the *non-accredited media* and is a community with mixed interests. The number of journalists who fit within this category has grown dramatically since the Sydney 2000 Olympic Games; it now rivals the number of accredited journalists who report on the sports program.

As was mentioned earlier, at the 2008 Beijing Games there were about 11,000 non-accredited media and 24,562 accredited media.[1] At the 2012 London Games the numbers were similar, but there were also more non-accredited media centers (NAMCs) in London, making total figures very difficult to estimate. In part, this is because the demographic of those communities is diverse and not uniformly counted. For example, non-accredited media may include a political correspondent from the *New York Times* or a freelance writer for a small specialist magazine. Alternatively, such journalists may be working freelance during the Games, or under contract to produce a series of documentaries for a large national broadcaster. They may be working for a rights-holding media organization, or for a rival company seeking to create Olympic content in order to piggyback on the public interest in the Games.

The second sense in which one may describe new media in the context of the Olympics has to do with the growth of social media and the proliferation of citizen journalism. This category spans a wider range of new-media artifacts, environments, and practices, from the micro-blogging environment of Twitter (on which accounts are set up by the Olympic organizing committee) to independent filmmakers who share their work via YouTube or Vimeo. A good example of this is a group of British Council filmmakers who produced work to be shown in Rio during the 2016 Olympic Games as a way of connecting the 2012 London Games legacy with the latest Olympic city. This second category is defined by its utilization of pervasive Internet-based technologies and, often, is interested in reporting what is happening in the world around them for political, civic, or personal reasons.

This chapter considers the first of these two categories, the new population of non-accredited journalists; while the following chapters focus on social media and citizen journalism. All these categories of journalists share the common distinction of *not* having access to the main media facilities at the Games, which is provided only to accredited persons, though it is also important to note that this community is likely to expand even further with the launch of the Olympic Channel. Already there is a range of creative media professionals at the Games who work to produce "legacy" material not aimed at live-television audiences. The Olympic Channel aspires to create Games-time content that may be added to broadcasts, and generating such content will require many more media professionals.

The Non-Accredited Olympic Media

Ever since the development of media accreditation at the Games, the International Olympic Committee has established guidelines that determine what it requires to be recognized as an Olympic journalist. First and foremost, it requires having an accreditation from the IOC or the Organising Committees for the Olympic Games (OCOG) during the Games. Recognition as accredited confers the entitlement to access aspects of the Olympic Games sports program. In general, this category comprises the rights-paying broadcasters and the print media, as was explained in the previous chapter. Their presence at the Games is predicated on their role as the storytellers of the Games. Their words, images, and audio artifacts are the basis for engaging audiences emotionally in what happens on the playing field. Commentators assist audiences in understanding both the context of an athlete's performance and also provide the most immediate collective emotional response to its significance.

In contrast, the non-accredited media do not focus on the sports events at all; at best, they focus on stories about what happens around the competition. This may include feature or human interest pieces on the local environment, but the key insight into what kinds of stories they tell comes from understanding that the non-accredited media center is funded through domestic, city-based support. The non-accredited media centers ensure that host cities are able to promote their interests, whereas Olympic media centers promote what happens within the Olympic competition venues.

The rise of the non-accredited media is due largely to the fact that host cities need to maximize their exposure during Games time and to manage the domestic press narrative during the Games. Yet the non-accredited media succeed largely because of the growing number of digital journalists, who make up a significant part of the NAMCs registered population. By studying the origins, the functions, and the development of the non-accredited media centers and their populations, one can understand more about change within journalism and how new narratives on the Olympics, as well as new conditions of media production are emerging. The non-accredited Olympic journalist population also highlights the challenges facing traditional media outlets. In the absence of IOC guidelines to manage such a population, the criteria for defining a journalist have been more fluid than for the accredited media.

The first organized attempt at coping with the growing number of journalists outside the official accredited list came during the Barcelona 1992 Games. The Barcelona City Council recognized the importance of using the Games as a platform for promoting the city and the region in order to attract and nourish the attention of media writers from non-Olympic-rights-holding organizations that would not have access to the sporting venues. Thus, it supported the creation of a center within the Barcelona Press Service. This center was organized in collaboration with the Autonomous University of Barcelona and focused its services on the specialist press and scholars interested in the history of Barcelona and Catalonia, and in particular the Catalan cultural identity. This experience was highly regarded by local authorities and encouraged subsequent Olympic host cities to do the same. However, the Barcelona center lacked visibility and relied on very limited technical and financial resources.

Over the years, the arrangements for non-accredited journalists have become more and more substantial, taking the form of specially constructed media centers. Non-accredited journalists are still required to go through an accreditation process to gain access to such facilities, but the process is not managed by the IOC or the OCOG; it is the city's own parallel process. Journalists who achieve accreditation through that process will not have access to Olympic competitions, but they may have access to any number of events that occur around the sports, such as the torch relay and the Cultural Olympiad.[2] There may also be related activities happening within the city that target these journalists specifically. For example, during the 2012 London Games the London Media Centre provided news about functions set up by the film industry or about West End shows. In addition, any number of sponsors may seek out non-accredited reporters, interested in the hope of promoting stories around new technology, or design, for instance.

While having a NAMC accreditation does not permit access to Olympic sport, having an Olympic accreditation sometimes confers automatic registration with the NAMC. In this sense, there is a kind of hierarchy in journalist status at the Games, a situation reinforced by the fact that even accredited journalists will have conditions on their freedom to roam, depending on what level of accreditation they have. Specifically, an infinity symbol on an accreditation will denote entitlement to access nearly everywhere, while some accreditations may be venue specific.[3] While generally speaking an infinity accreditation offers the most amount of freedom, reporters will have specific press accreditations that focus their access on sports venues, rather than on what is happening around the city. Yet at each Games, the character, population, and political importance of the new media has changed and subsequent sections offer some insight into how they have varied from one Games to the next. These components of what happens around the Olympic Games media infrastructure are a helpful reminder of one of the crucial transitions in recent media history toward democratization, but also an expansion of what counts as reporting and what is required to undertake such work at the Games. The new models of journalism that surround e-sports are products of this expansion and they describe a world where the media economy is less constrained to a small number of channel owners and where information is made much more available and reaches to audiences more directly. However, since the 1992 Barcelona Games, the NAMC has steadily evolved.

Sydney 2000: 7,000 New Olympic Journalists

After Barcelona 1992—and to some extent Atlanta 1996—the first major investment into provisions for the non-accredited media took place at the 2000 Sydney Games. By 2000, the commitment to such centers had been upgraded considerably. Sydney's main non-Olympic media facility to welcome the broad range of journalistic actors—accredited and non-accredited—was called the Sydney Media Centre and was situated in the fashionable downtown area of Darling Harbour. It was a collaborative effort by the Commonwealth Department of Foreign Affairs and Trade, the Australian Tourism Commission, Tourism New South Wales, the Department of State and Regional Development, and the Sydney Harbour Foreshore Authority. These organizations aimed to enhance the city and regional economic development via the promotion of its leisure and business tourism offerings. The Sydney Media Centre was also deemed necessary because the 1996 Atlanta Games had failed to accommodate this new population of journalists. An Australian parliamentary debate commented:

As Atlanta found to its cost, if … journalists are not looked after by being given good facilities from which to operate, if they are not provided with assistance in delivering interesting stories, the result is a deluge of media coverage critical of the city itself and critical of the Olympics preparations. We were absolutely determined that this would not happen in Sydney. (Murray 2000)

The establishment of the Sydney Media Centre was an attempt to promote local interests and a way of ensuring that journalists without access to the accredited venues had access to other facilities and stories. In short, it emerged as a result of a media management strategy and a facility-based way of encouraging a broader sense of what constituted the Olympics narrative and supplementing the work of the Main Press Centre (MPC) and the International Broadcast Centre (IBC). Located at the border of the harbor, the Centre provided filming locations for broadcasters and a spacious bar-restaurant in addition to the common provision of working and communication facilities, information stands, press releases, daily keynotes, press briefings, promotional events, and conferences. Its location also served as a base for accredited members of the media who required facilities closer to the city. (The Homebush Bay Olympic Park, where the main Olympic media facilities were located, was quite a way out of the city.) Several days before the start of the Games, the center had registered more than 3,000 media representatives. By the conclusion of the Games, 5,000 journalists had been registered at the Media Centre (Murray 2000, p. 9274). The venue hosted various high-profile events, including athletes' panels and press conferences with people who had take part in the opening ceremony. In this sense, there were the beginnings of an overlap between the official Olympic program and content programmed by wider stakeholders, outside of the main sponsorship arrangements. Indeed, the Sydney Media Center provided a vehicle through which lower level sponsors might be able to showcase their organizations, products, and services.

Salt Lake City 2002: The City's Media Center

In Salt Lake City, provision for the ever broader and technologically diverse non-accredited media was distributed between two different centers, each of which had different purposes, overseen by different organizations. The Utah Media Center was located in close proximity to the official Main Media Center in the heart of the city. It was an initiative of the Utah Travel Council with the support of the Chamber of Commerce and Visitors and Conventions Bureau in Salt Lake City. A second hub was created at the initiative of the Chamber of Commerce and was located in Park City, home of one of the most popular ski resorts in the area and a central point of access for a wide range of Olympic competition venues. The Utah Media

Center was the larger of the two. As in Sydney, it hosted high-profile events, including the only press conference given by Rudolph Giuliani, the mayor of New York City, who discussed the situation in the United States in the aftermath of the 9/11 attacks.

At the Salt Lake City Games, it was already apparent that the non-accredited media centers provided a crucial role for those media organizations whose accreditation allocation from the Organizing Committee or the IOC was below the number of people they wanted to bring to the Games. Furthermore, the more flexible security of the NAMC allowed greater freedom for reporters to undertake their work in a supportive environment. Yet it was also apparent that the investment from the city might have been far in excess of the day-to-day use of the venue. At the very least, the media centers had especially busy moments around important functions and events, but significantly greater periods of the days when they were almost empty and with very little activity taking place at all, raising questions about the return of the investment, or indeed the function of the space. The role of a press work room in a digital age is likely to come under further scrutiny as time goes on, and the NAMCs are a good example of this changing value.

Athens 2004: The Politicians' Media Center

For the 2004 Athens Games, the main NAMC was located in the Zappeion Center, directly next to the city's main square, Syntagma. In this case again, it is important to note that the Olympic Park was located some way out of Athens city center and so the NAMC provided a hub for journalists in the heart of the city. The Zappeion Press Center (ZPC) was established in a building that had historic value for both the city and the Olympic Movement, as it was the headquarters of the first Modern Olympic Games in 1896. This NAMC was far greater in size and political significance than its predecessor in Sydney, evidencing the growing centrality of the NAMC within the host city's Olympic delivery strategy. Furthermore, it became the site of more and more high-profile events. For instance, the day after the Athens 2004 opening ceremony, the ZPC hosted the formal signing of the Olympic Truce wall, an initiative that brought together heads of state, royalty, and IOC dignitaries, to symbolize their common commitment to the Olympic Truce. Notably, this took place outside of the normal, expected security requirements of Olympic venues and among the non-accredited journalists.

The ZPC also hosted a number of other symbolically important events, such as a presentation for the Melbourne 2006 Commonwealth Games and the presentation of the Cultural Olympiad. Each day, the ZPC held press briefings conducted by the Ministries of Public Order, Sports and Culture. Furthermore, there were opportunities for journalists to meet athlete celebrities, including Cathy Freeman, the Australian Aboriginal athlete who lit the Olympic cauldron in the 2000 Sydney Games and the city mayor. There were also daily events for journalists to attend, which, for the first time included some crossover with the Olympic sports program. Specifically, ZPC journalists were able to travel on a bus to Olympia and watch the shot put competition within the ancient site—the first time it had been used for competitive sport since the ancient Olympic Games. Again, this evidenced the growing closeness of the peripheral new-media populations with the accredited media. In this case, both communities shared the same journey down to Olympia, enjoying the same access.[4]

However, it was the political role of the ZPC that really defined its function during the Games and its importance as a daily press center for national journalists in particular was particularly crucial. Daily briefings covered such topics as the sale of Olympic tickets on the black market, questions about security, concerns about transportation, foreign policy conflicts during the Games, questions over eligibility of high-profile dignitaries from nations with controversial political leaders who sought to enter Greece, among others. Speaking after the Games, Greece's Secretary General of Information, Panos Livadas, emphasized the importance of this alternative media community:

[T]he dominant force that contributes to the formation, the reinforcement or the transformation of these stereotypes, are … not exactly the "accredited" media—that are present to report the sport and the athletics—but the non-accredited media, that "invade" the hosting country in order to report back the details of the hosting country, the society, the economy, the strong and the weak points that all our societies have. And, of course, to criticize. (Livadas 2005).

Outlining the investment from Greece to the ZPC, Livadas said that "40 press offices of our embassies around the world, made hundreds of meetings each, explaining the Zappeion concept for more than 6 months before August 2004" (ibid.). He also emphasized that the role of the provision was to operate efficiently in a crisis situation and, as a resource, there were 720 workstations with "three studio robotic cameras," along with outside broadcast vans and other editing suites. Together, these characteristics of the ZPC articulate its role as an important space for local politicians to reach journalists and as a contingency in the event of any unforeseen circumstances compromising other facilities.

Torino 2006: The Bloggers' Media Center

Winter and Summer Olympic hosts always look toward their respective predecessors, both as a reference point, but also often as a benchmark, above which they expect to deliver. Thus, comparing Torino 2006 with Salt Lake City 2002 shows a further increase of provision for the non-accredited media. The Torino Piemonte Media Center (TPMC) offered unprecedented facilities for journalists, including a vast and richly endowed press room with large-screen projections of athletic events, wireless computing, and gourmet regional cuisine. By 2006, the emergence of social media made Torino the first post–Web 2.0 media center. It had strong representation from online publishers and journalists, many of whom were bloggers. By this time, a number of bloggers had established enough publishing credibility for the organizers to look beyond traditional print and broadcast journalists in determining what efforts should be made to embrace these new journalists in official and quasi-official venues. Yet blogging was also novel enough to ensure that not very many were doing this in a way that was terribly successful in reaching large audiences. The range of bloggers at the TPMC included local as well as overseas writers, many from Vancouver, the next Winter Olympic Host city. At the TPMC, it was also apparent that foreign journalists were particularly welcome, regardless of the type of platform they represented and this reveals the wider aspiration of host cities to engage international journalists in particular, in an attempt to access new visitor markets.

In contrast with the Salt Lake City 2002 media center, the focus in Torino was not on promoting winter sports; it was on promoting the region of Piemonte, especially its culture, heritage, and cuisine. Moreover, one of the values of the center was predicated on Piemonte wanting to present an alternative impression of their identity, as a post-industrial region. Again, events staged at the center built on some of the Olympic assets that the IOC did not seek to protect during Games time. For instance, on one occasion, a press conference took place with the designer of the Olympic torch. Furthermore, there was a more unified communication system around the different accreditation processes, with the TPMC having a link from the main Torino 2006 organizing committee.[5] Though this may seem a minor point, it shows the increasing intimacy between the accredited and non-accredited media strategies. But another characteristic of the TPMC, which was becoming a common feature of all NAMCs, was a sense that the idea for producing such a venue was derived from city politics, rather than some transfer of knowledge system. While officials from Torino experienced the equivalent centers in previous Games, its establishment was not something that could be taken for granted, as might be said of the media centers that IOC requires host cities to construct. The NAMCs were not yet seen as crucial to the successful Games operation, but managing the expansion of the journalist population was becoming an additional challenge.

Beijing 2008: The Professionals' Media Center
Recognizing the role the NAMC would play in promoting the historical, cultural, and social elements of Beijing to the world, the organizing committee's Service Guide for Foreign Media Coverage of the Beijing Olympic Games and the Preparatory Period (Beijing Organising Committee for the Olympic Games 2007), took into account the needs of this population. In this document, the Beijing Organising Committee for the Olympic Games (BOCOG) expressed its intention to host an NAMC that would accommodate more than 10,000 journalists, including representatives from the more than 2,000 newspapers that exist in China, along with other international media. While this revealed further progress in the provisions made for non-accredited journalists, the implication of the growing closeness between the local government and the organizing committee were not necessarily positive, from the perspective of opening up to new-media publishers. After all, the IOC's management of the media operates within a tightly controlled structure and the increased visibility and integration of the NAMC with official structures was beginning to suggest that it would lead to the implementation of tighter restrictions on access and narrowing the range of participants it hosts. In other words, as more resources are diverted toward the non-accredited journalists, the rights-paying media may feel increasingly that their investment does not guarantee enough exclusivity. In this respect, further centralization may imply the development of greater

control and restrictions over these otherwise relatively free Olympic reporting spaces.

Nevertheless, with the expansion of non-accredited numbers, the consequence of such changes was greater and wider journalistic coverage of non-sporting elements and/or their integration within sports broadcasting may have been better as a result. At the time, the vice-director of BOCOG's media and communications department, Wang Hui, emphasized the diversity of media coverage during the Olympics:

[M]edia are concerned not only about who won a gold medal and set a world record during the Olympics, but also about the Olympics hosting country's landscape, the hosting city's characteristics, local people's lives, how they participate in the Olympics.

However, one of the more important developments of Beijing's Games was that the organizing committee emphasized that the NAMC would host only *professional* journalists, specifically those who did not have access to the MPC and the IBC. This was the first time an NAMC required an applicant to demonstrate a professional journalist credential as a condition of entry. (In the past, it was sufficient to demonstrate a presence within a media outlet.) In consequence, this meant that many freelance journalists were unable to access Beijing's non-accredited media center. Yet with the rise of the Internet population in China (see China Internet Network Information Center 2007) it was unlikely that many of these non-accredited journalists would be either professional in the sense of their having a national press card or an assignment letter from a media outlet. In the past, considerably less rigor had been applied to applications from journalists while Games were in progress, as local authorities were pushing to attract publicity about non-Olympic-related causes in international outlets. In China, this imperative was not apparent. Furthermore, as the Games began, the rules about a 15-day advance notice on applications for accreditation became unnecessary; they could be processed on the day of arrival. In part, this seemed a direct consequence of the media centers being much less busy than anticipated. However, for Beijing's center this was not the case. To enter China as a journalist required a special visa, which formed part of the access application to the NAMC. This obstacle meant that many freelance or semi-professional reporters visiting Beijing to cover the Games weren't able to enter the NAMC at all.

Common Features of Non-Accredited Media Centers

The period from 1992 to 2008 marks a distinct phase in the life of the NAMCs, as they steadily became consolidated as an established media venue within the Olympic program, increasingly professionalized and where access is now almost certainly conditional on applicants demonstrating a credible journalist affiliation. Both the Vancouver Games and the London Games created NAMCs that were similarly large scale and with considerable investment from the city. Rio de Janeiro did something similar with its Rio Media Center, hosting around 8,000 registrants to the city during Games time (Rio Organizing Committee for the Olympic Games 2016). Vancouver's British Columbia International Media Centre (BCIMC) was located in the most central part of the city—Robson Square—and was a hive of activity. For London 2012, the London Media Centre (LMC) was also located in a central venue, in the heart of Westminster. Equally, the Sochi Media Centre in 2014 operated under similar principles. Questions about the trajectory of these entities within the Olympic infrastructure remain, but there are reasons to believe the NAMC is becoming even more centralized, as the venue becomes a more important asset for a city. For instance, the primary partner of the LMC was the credit card company Visa, also a Worldwide Partner of the IOC. This section scrutinizes these developments, identifying common shifts in what is taking place and what it means for debates about the disruptive potential of new-media platforms.

Over the years, some common features have emerged to distinguish the NAMCs from other media structures at the Games. First, they are physically and structurally separate from the major accredited media venues. The latter tend to be very close to the Olympic sports venues, whereas the NAMCs tends to occupy a prime city center location. In addition, the arrangements for non-accredited journalists tend to be established by the government of the host city and affiliated authorities, rather than the OCOG or the IOC. Because of this, the focus of these centers has generally been on the promotion of the local cultural milieu, with an emphasis on tourism and business opportunities, rather than sports (though their atriums or work rooms often have large screens displaying broadcasted competitions). Also, due to their greater flexibility in the registration of users, the non-accredited centers attract a more diverse range of journalists than are present at the accredited facilities, many of whom are not associated with mainstream media groups. (I use the word *diversity* here to describe types of media outlet and the range of subjects that they seek to cover, rather than appeal to the numbers of outlets or countries that are present in each.)

However, these venues are not specifically designed to function as "alternative" or independent media centers (Lenskyj 2002; Neilson 2002) the role of which would be to facilitate "the organization of (publicly advertised) Olympic-related protest events" (Lenskyj 2002, p.166) organized by "a diverse collective of media activists" (ibid., p. 167). Although the NAMC may include individuals with anti-Olympic inclinations, they are not established for that purpose. It is also unclear that any explicit anti-Olympic media producers have either sought access to or desire to be accredited by a NAMC. Rather, following the success of Sydney, the term *non-accredited media center* has been adopted at every Games since to designated an additional, mostly supportive journalistic community. Despite having been developed outside the official Olympic regulations, the NAMC have structures and functions that reveal significant commonalities, which become clear when examining their respective journalist demographics; the characteristics of location, facilities and stories; and the evolution of an ever-closer relationship with the host city's Olympic Organizing Committee.

In contrast to accredited journalists, most of whom represent mainstream media groups, individuals and companies registered at the NAMC represent a wide variety of organizations, including small outlets such as specialist culture and trade magazines, or community radio stations, for instance. Furthermore, those journalists who register at the non-accredited media center are neither regularly accredited in their own countries nor always professionally trained. Thus, they bring a variety of agendas, demands, experiences, and interests to these centers. Among those who use these facilities are people in the following categories:

1. official Olympic accredited journalists who find the location, facilities and environment more convenient or find the NAMC program of events to be newsworthy
2. journalists from IOC-accredited media organizations who do not have their own accreditation to the Main Press Center or International Broadcast Center, due to the limited quotas.
3. journalists from mass media organizations who do not have official Olympic accreditations
4. specialist press reporters and freelancers

5. journalists who run their own publishing outlets
6. professional online publishers whose work in online platforms is insep-
   arable from their personal online-profile as creative practitioners
7. unpaid "citizen" journalists interested in exploring and portraying
   alternative impressions of the Games.

Typically, the first four types can be characterized as professional journalists, whereas the final three categories tend less to be professionally trained
reporters. The last two categories are growing in numbers rapidly. In Torino,
video bloggers (vloggers) were plentiful for the first time. Notably, an
increasing number of journalists from categories 1 and 2 are using the non-
accredited facilities, working within the same environment as journalists
from categories 3–7, who were originally the targeted users. The wide variety of individual backgrounds and the unique situation of all of these journalists sharing the same facilities and attending the same conferences over a
concentrated period of time offer unexpected opportunities for interactions
that can lead to quite unusual collaborations. For instance, the agenda of a
press conference may be radically transformed simply because the interests
of the small press are different from the main stream large media; these
interactions also raise the possibility of the minor press' capability to influence and transform the agenda of established, mainstream journalists.

The NAMC's city center location places it close to relevant cultural
attractions and political institutions. Furthermore, content presented at the
NAMC emphasizes hospitality as much as political communication. It is a
facility in which those who underwrite it—local and regional authorities
as well as corporations and governments—are cajoling as much as hosting,
trying to extend the field of vision rather than simply provide access. As a
result, the NAMC retains a strong local character, which contrasts sharply
with the standardized framework of the MPC and the IBC, where facilities
present almost identical features from one Games edition to the next and
where stories typically exclude any social, cultural and political aspects of
the local host experience.

The NAMC has no consistent link to the accredited centers but is
increasingly being used to ensure representation of the host city's Organizing Committee in cultural, educational, and environmental matters. For
example, one feature that has been integral to the NAMCs since Sydney
2000 is the presence of the Cultural Olympiad, which has only a marginal presence at the Main Media Centers. The NAMCs have also become

hosts of high-profile Olympic-related events and are sites for information about popular Olympic features such as the Medals Plaza during the Winter Games and the LiveSites! None of these would feature prominently at the accredited media venues and, in some sense, the growth of the NAMC reflects the transition that the Olympic program is undergoing from being exclusively about sports competitions to becoming a city-wide festival during the Games.

Institutionalizing the New (Olympic) Media

A number of distinctions are necessary to make between journalists in the context of the non-accredited Olympic media. First, it is useful to provide a wider re-conceptualization of the Olympic media, which can be broken down into three primary categories:

• accredited (those who have accreditation to access Olympic Venues and the official IOC media centers, which are overseen by the IOC and OCOG)
• non-accredited (those who receive accreditation from the NAMCs, which is overseen by the host city's government)
• unaccredited (those at the Independent or Alternative Media Centers, as well as those acting as "citizen journalists").

Second, it is necessary to consider the range of ways in which each of these types of media contribute to or detract from the primary Olympic narratives. Thus, the accredited facilities communicate the official IOC and OCOG narratives; the non-accredited media centers offer an additional city or regional governance narrative, which has the potential to supplement or compete with the coverage of the former. Though one might expect that the IOC, the OCOG, and the host city's government would work toward similar goals, in practice they all are competing for different kinds of (positive) narratives and different kinds of media attention. For the IOC, the Olympic Games is an opportunity to showcase and reinforce the Olympic brand as a global entity. For institutions associated with the host city's governance structure, the Games offer an opportunity to showcase the local values and heritage, perhaps to boost the opportunity for global economic investment or tourism and to promote national pride. For the unaccredited media, these aspirations are much more diverse and can range from promoting specific kinds of political ideals to providing opportunities for engagement within a local community.

A third process associated with these media populations is their varying role in both institutionalizing and destabilizing the infrastructure of Olympic media production. Thus, the NAMCs are agents of institutionalization insofar as they are used to manage journalists who are external to the Olympic accreditation process and who might, as a result of not being managed, have an influential role in communicating negatively about the Games. However, as a result of this process of institutionalization, the NAMCs also are at risk as city-led initiatives, since their success can become a conflict for the exclusive, rights-holding media.

As the most recent, established new journalist population at the Games, the NAMCs have challenged how media coverage of the Games takes place and, as a result, what the Games mean to nations and people. To the extent that the Games aspire to be a publicly shared media event, the NAMCs are a crucial part of what can be called Sport 2.0, as it describes how media access has become democratized at the Olympic Games. Even without knowing much about what is actually published or broadcast by this community, it is clear that, for such media, the Olympic Games are not of interest for their sporting value. Rather, they are of interest because they provide an opportunity for intense formal and informal cultural and political presentations and positioning.

The Olympic Fringe Media

The presence of non-accredited media has grown progressively at recent Games and is distinct from the accredited media in one crucial characteristic: There is no limit on the number of journalists who can report from the NAMC, other than what facilities can accommodate. Additionally, the demographic of these reporters remains transient. It is constituted by a flow of accredited journalists, as well as reporters from non-accredited print, broadcast and online, as well as being populated by a range of non-professional, freelance journalists. The organizational infrastructure for non-accredited media is also loosely defined, occupying no place in the formal transfer of knowledge that takes place from one Games to the next. Instead, it is invented seemingly by circumstance, but increasingly by subsequent Games witnessing the facilities of earlier Games. As these grow, the impact and incentive to do the same has grown. In this sense, growing the non-accredited media population appears to be a positive value for an organizing city, but of little interest to an Organizing Committee. However, there should be scope for co-producing such spaces to maximize common interests.

The rise of non-accredited media at the Olympics presents a challenging set of circumstances for future Games and for the organization of mega-event media generally. The new populations of online journalists that were apparent at the 2006 Torino Winter Games revealed, for the first time, the capabilities of low-budget journalism. The needs of such reporters are different from those of professional journalists, as are their political interests. For instance, some of the Torino journalists felt constrained by the concept of having a physical media center—a venue—which created a restriction on movement during the chaotic period when the Games are in progress. Figuring out what kind of provision is necessary when journalists no longer rely on fixed broadband or power sockets is one of the new challenges for the hosts of the NAMCs. The only answer is for the Center to become an event venue in its own right, worthy of news coverage.

There is also a challenge for the organizers to determine how to provide accreditation to online journalists, many of whom would not fit into traditional definitions of what it means to be a journalist, such as holding a national union card. Many of the journalists at NAMCs since Torino were neither professionally accredited reporters nor professionally freelance reporters. Rather, a considerable amount of them had extensive followings of viewers of their blogs or online magazines.

The presence of non-accredited and new-media reporters at the Games reveals how the established model of understanding the Olympic media is in transition. While this may create logistical challenges for organizers, the expansion of the journalist category is consistent with the Olympic ideals, as it refocuses attention on citizenship rather than on corporate media interests as the unit of what makes the media important. However, by the same token, non-accredited media challenge the financial structure of the Olympic Movement, which is reliant upon the sale of intellectual property (i.e., broadcasting rights). If the NAMC continues to grow in prestige and influence, one can foresee Olympic sponsors' seeking to curtail or absorb its function. Alternatively, the pressure to manage media narratives on the Games might lead to its abolition through a contractual stipulation between the IOC and the host city.

Future hosts of the Olympic Games will benefit from exploring how to harness the role of the non-accredited journalists and from considering strategically how best to furnish them with opportunities to tell stories about the Games. Although the political focus of media relations at the Olympics is on the period during which the Games are taking place, many of the non-accredited journalists are working on features that will be published soon after the Games, but which are not competing for publication space and which don't have pressing deadlines. The NAMCs provide a dynamic setting and can deal with queries in a more flexible way than the official accredited centers. Most important, they offer alternative stories that allow a greater understanding of the Olympic host and can ensure a fairer representation of local communities.

NAMCs provide a mixed environment at the Olympics, a place where different media agendas and populations converge. They are regulated, but they are not official Olympic venues, as the absence of the Olympic rings and the absence the word *Olympic* within a center's name indicate. Thus, they are not subject to the strict level of regulation of the Olympic venues. This distinction is important when considering their role in the creation or definition of additional narratives. Moreover, it informs our understanding of what character media coverage of the Games might exhibit. While independent or alternative media centers are sometimes explicitly anti-Olympic, this chapter has focused on the *supportive* media spaces within the Olympic city. In part, this is because there are opportunities within the present Olympic system to challenge the dominant media structures, as a result of these developments. Thee more relaxed accreditation process of the NAMCs, along with the governmental oversight and capacity to gain access for media to important political and cultural events, provides a rich set of circumstances through which the highly regulated media structures at the Games can be circumvented.

8 Social Media and the Olympics

While the rise of the non-accredited Olympic journalist has expanded the range of coverage of the Olympic program, the growth of social media has transformed the range of locations in which Olympic media content can be found. And although these two trajectories have been independent, they have also been mutually reinforcing. This chapter focuses on the emergence of social media as a new kind of Olympic journalism that is highly reliant on the co-development of content and audience involvement. Yet the crucial determinant of its success is located in a wider crisis within the media industries over legitimacy, authority, and independence. In this respect, it is fitting that the opening ceremony of the 2012 London Games featured the inventor of the World Wide Web, Tim Berners-Lee, and his classic phrase "This is for everyone." That sentiment helps to explain the power of social media in the syndication of Olympic news.

Since the creation of the Web, there has been a significant amount of change in users' experiences, but also in how content is moved around digitally. The era of social media is demarcated by the idea of Web 2.0, which was popularized in 2004 via an O'Reilly Net conference to describe a new working ethos around the Web and a redefinition of the technical means by which content could be shared. Though some would argue that the expression *Web 2.0* simply denoted a different set of architectural principles for how content is created and moved around the Internet, the idea of an ethos is a more compelling way of making sense of how it differs from Web 1.0. Even the idea of sharing may be evidence of this change in what people expect from media content, moving away from an era in which information is broadcast in a unidirectional manner to a sense in which the content can belong to everyone.

The underlying idea of Web 2.0 defined the era of social media and generated new conversations about what the Web should be doing for people, which included making information more portable, more creative, and more visual. Subsequently, *2.0* would quickly become a metaphor for innovation, alternative ways of thinking, and new forms of human development, drawing on the terminology used to describe new versions of software. Barassi and Trere argue that its popularization was "profoundly influenced by business rhetoric" (2012, p. 1282). Yet its use in the past ten years to describe a paradigm shift in digital design remains useful as a way of thinking about what changed for Web users. Perhaps the most powerful application of the "2.0" era is found in Steve Fuller's 2013 book *Humanity 2.0*, which testifies to the changing conditions of human evolution, a theme brought about at least partially by increases in computing power. The term *Web 2.0* has flourished as a result of progress in computer technology, the constancy of Moore's Law, and the ongoing pursuit of Ray Kurzweil's "singularity"—the moment when the intelligence of computers overtakes that of humans. Though much of the Web 2.0 entrepreneur spirit was only tangentially connected to these broader debates about the development of humanity, it also signaled a radical overhaul of our knowledge economy, emphasizing a way of thinking about communication and social order. For instance, Wikipedia has its roots in the Web 2.0 era—an era defined by the ability for many people to have the role of author and where the idea of collective wisdom replaces individual expertise.

Web 2.0 was a new kind of World Wide Web, distinguished by new forms of publishing, people, places, and processes. These new environments were far superior in their ability to connect people, would ease consumption and production, and would transform how people mobilize digital resources for a range of creative, political, economic, or social uses. The concept of Web 2.0 was also a symbol for a better Web, one more equipped to deal with a burgeoning mobile culture. Yet it was also a place where debates about how intellectual property rights would be marshaled in the future became ever more complex. Websites were now designed around the concepts of syndication, sharing, and re-publishing, and brand identities were easy to hijack on a scale that was previously unimaginable. The only limiting factors were bandwidth and the cost of accessing the Web outside of Wi-Fi range. Some of the best-known platforms emerged around that time, including Facebook and Wordpress, two of the most popular platforms for sharing content across the Web. In the next two years, YouTube and Twitter were launched.

These changes happened around the time of the preparations for the 2006 Torino Games—only ten years after the Atlanta Games, the first Olympic Games to have a website. The consequences of Web 2.0 for Olympic organizers was a vastly expanded publishing and broadcasting community. Moreover, almost overnight a vastly larger population of publics had the means by which to publish their own content. Spectators could record their own Olympic sports content and broadcast it on the Web in real time, albeit illegally. As was noted earlier, Torino 2006 was also the first Games at which the non-accredited media center began to host bloggers. Two years earlier, in Athens, this was not yet possible; now it is hard to imagine an Olympic Games in which news doesn't circulate through social media.

These changes would quickly set the tone for the Olympic movement's engagement with its online public, transforming methods of communication and developing new audiences through the growth of social media. By the time of the 2010 Vancouver Games, there were two parallel processes taking place around the uptake of new digital media, both intimately connected to the tools afforded by social media and the new architecture that was Web 2.0. The first of these trends was the growth of *user-generated content*, the fruit of social-media labor. The second was the growth of *citizen journalism*, which became increasingly tied to the rise of social-media environments. This chapter deals principally with the former of these trends, the shift in digital communications toward *social media*—a term that has now become ubiquitous in discussions about the Internet, but which remains essentially contested as a media category. It considers how social media have affected the infrastructure of the Olympic Games, along with the individuals who are essential to its production—officials, athletes, sponsors, and the media outlets that seek to leverage interest in the Games to boost their numbers of viewers, listeners, or readers.

Web 2.0 and social media must be seen as two sides of the same coin. In order for media to be social in the sense implied by the term *social media*, a set of communication protocols that is qualitatively different from the Web 1.0 version of the Internet is necessary. Thus, Web 2.0 websites were built on extended mark-up language (XML), which was coded in such a way as to enable sharing, re-publishing, and embedding almost infinitely and with minimal labor—sometimes automatically, as may be said of Facebook's recent Instant Articles initiative. As Sweney (2015) notes,

Currently, mobile readers have to click on a link in their news feed with a wait of more than eight seconds for the article to load in another Web page—a slow experience in the fast-paced Internet world. ... The new initiative, called Instant Articles, will see stories run within Facebook that the company says will make for a seamless loading experience 10 times faster than the current system (online).

Among the earliest adopters of Facebook's Instant Articles were newspapers (including the *New York Times*, the *Washington Post*, the *Guardian*, and the *Daily Mail*) and broadcasters (including NBC News and the BBC). Facebook's Instant Articles feature exemplifies the principles of the Web 2.0 ethos, which includes allowing content to move around seamlessly for the end user and giving content a life far beyond its primary publisher's platform. In fact, the Instant Articles by third-party publishers stay within the Facebook environment, which makes it easier for users to scroll through their Facebook feeds more easily. Although the status of Web 2.0 as a distinct paradigm is disputed, its pertinence to the print media and journalism cannot be overstated. Yet the operating architecture of Web 2.0 has been present since the Web 1.0 days, when Amazon, Trip Adviser, and eBay were pioneers of a coming Web 2.0 revolution. Furthermore, its capacity to transform the conditions of digital participation has been questioned in cultural studies. For instance, one of the most contested debates in digital cultural studies concerns whether the proliferation of social-media technologies exacerbates the problematic consequences of consumerism or whether it revitalizes civic life and individual agency. After a decade of social media, it seems reasonable to conclude that they do both. Social media thus offer a way for companies to make their marketing and sales techniques more precise in order to sell users more things, but also offer a way for users to demonstrate their social concerns more effectively.

Considerable rhetoric around Web 2.0 culture draws on the idea that what distinguishes our present media culture is how it has transformed people into *prosumers*, the antidote to consumer culture and a concept that gives primacy to the way that consumers become part of the means of production—for instance, by contributing to open-source software development or by editing Wikipedia entries. More broadly, digital engagement and participation has shifted from being a process that is largely about distributing content from one to many to a many-to-many configuration. In this new model of communication, the ability of individual narrowcasters to reach large audiences and for their messages to be syndicated across multiple networks is enabled by the long tail of each individual user, which reaches more and more individual hyper-local communities. Each user is a link in the communication chain, which allows ideas to spread. In this respect, Web 2.0 and social media are seen as mechanisms through which individuals can reassert their citizenship as journalists and, perhaps, as social watchdogs.

Admittedly, this kind of rhetoric sounds remarkably similar to the claims made about Web 1.0 in the 1990s, which held up the Internet as the panacea for society. After all, the act of sharing another organization's content may also be described as a form of emotional labor, an act which has the effect of working for an organization by allowing their intellectual property to be distributed further. In this respect, sharing content on social media is an act of taking part in a large marketing project. Nevertheless, the ease at which it became possible to publish ideas across platforms was a significant change in the Web's publishing architecture.

One of the biggest differences between Web 1.0 and 2.0 was that creating content in Web 2.0 environments no longer required vast technical knowledge or any training at all. It was no longer necessary to learn a programming language to create websites, nor was it necessary to learn how to design or use file-transfer protocols. Publishing in a Web 2.0 era also required very little knowledge of how to make websites searchable, and no longer was it very difficult to create highly visual content. All these tasks were effectively de-skilled through new social-media platforms, the use of which required little more than the ability to navigate a Web browser. This is not to say that being successful online was devoid of strategy or insights; it is to say that publishing content out into the world had become much simpler. This ease of publishing, along with greater accessibility to the Web, is perhaps the most crucial difference between the two eras.

Yet, despite the expansion of creative freedom that the Web 2.0 era has brought, there is also a sense in which they enslave users to a relatively few large (new) media giants, which are commercially exploiting the free labor that underpins the success of their platforms. For example, when Facebook users update their status, upload content, or search for information, they contribute to the vast database of insight that is subsequently monetized by Facebook. In this sense, the expansion of our sense of freedom is accompanied by the exploitation of our play by organizations that seek to sell us more things by knowing more precisely what we value. Boyle and Haynes draw attention to this when considering soccer in the age of new media. They claim that digital media "marks another staging post on the long road of commercialisation which the media and football industries have been embarked on since their first meeting" (2004, p. 161). Indeed, they argued that this trajectory led to a loss of citizenship, and that the global reach of digital culture was detrimental to more traditional and enduring local connections among soccer fans.

Much about digital culture has changed since Boyle and Haynes published their analysis. The Web 1.0 period was characterized by monopolization of the Web by such giants as Google and Yahoo, which matured in that period. Yet the rise of user-generated content changes the dynamic in various ways and, despite the fact that there remain financial models that allow such new-media giants to maintain their dominant position, the relationship between these companies and the user has changed. A simple example of this is found in the way that somebody is, in a Web 2.0 era, able to use freely available platforms to publish content in a variety of forms, where Web 1.0 was much more reliant on the purchase of software and space online to be a competent website publisher. The existence of Google and similar companies has brought about the demise of companies whose software only a few people could afford. Even the creative tools provided by Adobe—which have been industry standards for some time—are now available by a subscription rather than for one one-time payment, which allows customers to always have the latest version of software and to pay less per year. The considerable implications of Web 2.0 for the Olympic movement have been demonstrated by the plethora of Olympic digital artifacts available today. With Facebook, Twitter, Instagram, and Snapchat generating new communities that bypass the need for Google's search facility, even the dominance of Google is now in question.

A wider ability to produce Olympic media content across new platforms brought with it the possibility of engaging audiences differently, rather than just treating them as spectators or viewers of content. Yet it also enabled more people to re-purpose material in ways that may be considered as a breach intellectual property laws. For instance, a spectator at an Olympic event who took a close-up video of an athlete receiving a medal at the Games and then uploaded it to YouTube would be in breach of conditions specified on his or her ticket. Yet it is plausible that many spectators would neither know nor care about this infringement, and indeed that they might not even perceive the act as an infringement. Indeed, one may ask whether it would really matter if someone did this, were it not for the fact that the content may spread to locations beyond those they had intended. These kinds of challenges became more urgent in a Web 2.0 era—in part because, even if the user does not make money, one can argue that the value is derived by the platform, which hosts the content. The platform draws the attention of new viewers and may even be able to monetize space around the content, through advertising. However, it is far from clear that such acts would threaten the Olympic Games. After all, the International Olympic Committee has increased its income throughout the Web 2.0 period. Rather than diminish television viewing, social media seem to encourage more of it. This supports the claim that, far from challenging old systems and bringing radical change, new media help to reinforce established capitalist tendencies within elite sport (Dart 2012). Yet the IOC has sought to differentiate what happens in social media from other media environments, as the next section explains.

The IOC's Blogging Guidelines

The IOC's first "blogging guidelines" for "accredited persons"[1] were released in February 2008 and were an attempt to ensure control over the distribution of media content about the Games. (See International Olympic Committee 2008.) The rationale for such guidelines has been much debated, but their purpose is broadly to protect Olympic stakeholders' interests, either by directing accredited persons away from unknowingly publishing content on unauthorized commercial sites or by prohibiting "ambush marketing" (that is, willful exploitation of a private contractual relationship that might undermine the exclusive association rights of the official Olympic partners).

Ambush marketing has become commonplace at the Games. For instance, in 1992 the American basketball player Michael Jordan famously covered up the logo of his team's sponsor, Reebok, when receiving his medal, so as to not compromise his personal sponsorship contract with Nike. In this sense, the blogging guidelines extend the protection of the Olympic brand assets, which underpin the financial viability of the Games. However, the Web 2.0 version of ambush marketing is rather more nuanced, as Nick Symmonds' Twitter tattoo demonstrates. More will be said later about how ambush marketing takes place in a Web 2.0 era. First, I want to focus in more detail on the IOC blogging guidelines, since they influence the conditions of Olympic social-media life.

There are various aspects of the IOC's 2008 blogging guidelines that are relevant to consider when attempting to understand how Olympic media content is overseen by the Olympic industry. I use the word *industry* here to refer to the totality of the Olympic stakeholders—not just what the IOC designates as its Olympic Family, but the range of actors who are legally invested in the Olympic interests and who have a stake in maintaining brand exclusivity. When comparing the initial blogging guidelines with the 2012 version, there are important differences that reveal how the cultural value associated with social media has increased. According to the IOC's 2008 blogging guidelines, blogging is "a legitimate form of personal expression and not ... a form of journalism." The guidelines go on to outline a series of conditions that were intended to guide the activities of accredited persons at the Games. The decision to regard blogging as a form of personal expression, rather than journalism, is the primary reason to interpret the guidelines as reasonably liberal rather than deliberately restrictive.[2] Though this view may be challenged by staunch supporters of Web freedom, there are contextual reasons to reinforce this interpretation of the guidelines.

The first reason is that the guidelines apply only to people who have entered into a contractual relationship with the IOC, either personally or indirectly through an organization that has such a relationship. This category encompasses athletes, officials, coaches, and, generally, anyone with official Olympic accreditation. The guidelines do not attempt to restrict the blogging freedom of the general public or how people may seek to publish content online. An Olympic ticket holder who endeavors to commercialize media content from the spectator experience, or even to upload sport footage to a free platform, will find that the small print on his or her ticket states a similar restriction. For instance, tickets to the 2012 London Games stated among the terms and conditions of ticket purchase the following:

You agree that any images, videos, or sound recordings of the Games taken by you may only be used for private and domestic purposes and cannot be used for any commercial purposes, whether on the Internet or otherwise.

Such conditions are consistent with what takes place at other cultural events—such as filming in cinemas or at concerts, so one ought not get too carried away with claiming them to be unreasonable impositions. Nevertheless, social media test the limits of these stipulations, since it isn't clear whether uploading a photo of a playing field to a free publishing platform constitutes a breach of them. After all, while the platform may be free, the content is still located within a commercial space. It would have been far more restrictive if the IOC had defined blogging as a form of journalism, since the IOC's contractual relationships with broadcasters required it to monitor Olympic journalism.

Second, there is reason to presume that the IOC seeks to distinguish between the professional practice of journalism and the kinds of reporting activities that the general public may undertake. The latter are not subject to the kinds of expectations of professional journalism. For example, professional journalists are expected to maintain a certain code of ethics in the conduct of journalism, but the general public is not (even if it should). Of course, the salience of this distinction is at the heart of the debate about how media culture is changing. In recent years, criticisms of journalists' ethics and evidence of critical, responsible journalism by citizens raises questions about whether the two are becoming more closely aligned—in some instances, citizen journalists show more ethical concern and awareness than professional journalists. Yet in the context of the Olympics, it is still unclear what counts as either journalism or some other kind of news syndication.

Consider an athlete who has taken a video camera or a mobile phone to the closing ceremony of the 2012 London Games and live-streamed what he was seeing directly to a free-to-use commercial platform that had no rights associated with the Games. Such an act would have violated the IOC's guidelines in two ways. First, it would have been reasonable to argue that the athlete was making a personal diary of his experience and so would have been within the scope of what the IOC permitted. However, if the athlete's camera had captured something controversial, the footage might have created a whirlwind of media attention and, by implication, would have become material for journalism. Even in the absence of any controversy, the eye-level view of an athlete in the stadium among all the stars could have been exploited as premium media content, perhaps capturing more interest than images caught by rights-holding cameras, which would have been much farther from what was happening in the stadium. Second, commercial conflicts may have arisen as a result of such posts. Even if the platform used by the athlete had been free and, therefore, there would have been no commercial exchange between the athlete and the website, its existence would have been underpinned by a financial model—someone or some company would have been accumulating value in kind or in cash as a result of traffic generated by the content.

Thus, even publishing to free-to-use platforms presents a challenge for the IOC, which in the hypothetical case mentioned above would have had contracts for online content with a primary media partner. Moreover, a free platform used through the athlete's personal account might have had substantial non-Olympic advertising, which would have conflicted with IOC contracts. During the 2012 London Games, projects associated with the Cultural Olympiad uploaded content to their own YouTube channels, the videos of which were preceded by commercials determined by YouTube rather than the IOC. Such commercial associations have, as yet, no relationship to the IOC marketing model and thus may be described as a passive form of ambush marketing.

The IOC's 2012 blogging guidelines were similar to its 2008 guidelines in that two of the main concerns were brand protection and ensuring that any video footage from Olympic sports would not be transmitted by unauthorized accredited persons. Yet there were also some big differences. The 2012 guidelines began by emphasizing the importance of social-media *freedom*. Furthermore, a desire to distinguish between diary-like personal blogging and journalism was evident in the following passage:

The IOC encourages participants and other accredited persons to post comments on social media platforms or websites and tweet during the Olympic Games, and it is entirely acceptable for a participant or any other accredited person to do a personal posting, blog or tweet. However, any such postings, blogs or tweets should be in a first-person, diary-type format and should not be in the role of a journalist—i.e. they must not report on competition or comment on the activities of other participants or accredited persons, or disclose any information which is confidential or private in relation to any other person or organisation. A tweet is regarded in this respect as a short blog and the same guidelines are in effect, again, in first-person, diary-type format. (International Olympic Committee 2012)

The elevation of the third person as the mode by which journalism distinguishes its professionalism may seem simplistic—many forms of journalism are written in the first person. However, this seems to be the principle distinction between reporting and making a diary, as articulated by the guidelines. Yet there remain a number of cases in which the ambiguity of these guidelines could lead to situations of uncertainty. For instance, if an athlete posted a link to her main sponsors' website from a tweet in their personal Twitter account, would this be considered a breach of the guidelines, even if that tweet did not reference the Olympics? In short, do the social-media identities of all athletes become Olympic property during the Games period, or would only communications pertaining to the Olympics be covered by the guidelines? A new explanation is offered in the answers to frequently asked questions about the Rio 2016 guidelines:

As a principle, accredited persons should only use social media during the period of the Games for the purposes of sharing their experiences and communicating with their friends, family and supporters and not for commercial and/or advertising purposes. Accredited persons may only post about their sponsors, promote any brand, product or service on social or digital media or otherwise use social and digital media in a manner that creates or implies any association between the Games or the IOC and a third party, or its products and services, if they have obtained the prior written approval of the IOC or their National Olympic Committee. (International Olympic Committee 2015a)

Effectively, the IOC would seem to require that athletes think of themselves as ambassadors for the Olympic movement in everything they do during the Olympic Games period, rather than to use this period as an opportunity to advance their individual financial interests. Though the IOC may not be able to restrict such expressions legally—beyond what is covered in the Olympic Charter—the guidelines indicate that athletes are morally bound to limiting their positioning other commercial interests during the Games. While this may seem excessively restrictive, it also seems reasonable to claim that the attention afforded to any Olympic athlete during the Games is a consequence of an athlete's being a part of the Games and thus that any activity that undermines the Games' brand would be in conflict with that agreement, even if the content was generated outside of the Olympic venues. When comparing the 2014 and 2016 guidelines, there is a shift from a policy under which athletes are forbidden from posting commercial links to sponsors to a situation in which they must obtain the IOC's approval before doing so, suggesting more freedom to negotiate such interests.

The guidelines for the 2016 Rio de Janeiro Games make additional, important modifications to those for the 2012 London Games. For instance, during Rio 2016 certain areas of the Olympic Village are designated "no picture areas" to allow athletes some privacy—a practice that was introduced at Sochi in 2014. The revised guidelines maintain a clear distinction between freedom to post still photography versus audio or video content, the latter of which is much more restricted to personal use and no sharing on social media. A final, minor but revealing shift in the guidelines from their conception to the latest version is a shift in language from simply "blogging guidelines" to "social and digital media guidelines," suggesting further how the range of digital communications has expanded over the years.

When wrestling with the complexity of these cases, it is important not to conclude that social media are exceptional, as many such potential breaches could arise through other forms of activity. For instance, if an Olympic champion thanks a non-Olympic personal sponsor when being interviewed by an Olympic broadcaster, they may have made similar breaches of guidelines. Again, looking at the case of Nick Symmonds and his Twitter tattoo sponsor, it is difficult to conclude what kind of objections might surround such an act. In this case, the auction winner would have its Twitter account tattooed on the athlete's arm in return for sponsorship to the amount of the winning bid. The implication of this proposal is that the exposure of the Twitter account via the athlete's body would provide valuable advertising space for the auction's winner. Yet the case was something of a red herring in terms of testing the limits of Olympic ambush marketing, as the athlete admitted that he would be required to cover up the tattoo during any Olympic races because it would be seen as a form of unauthorized advertising. Nevertheless, Symmonds notes that covering up the tattoo with tape each time he runs is a reminder for audiences of the rules that restrict athletes from securing their own sponsorship deals), an act that alludes to Naomi Klein's "No Logo" thesis. Such a novel example of how social media may change the monetary relationships between athletes and the sports industries shows how complex the management of a brand has become in the era of social media. It also alludes to a future in which every part of an athlete's body can be monetized with digital advertising.

Despite the benefits of the IOC's definition of blogging as personal expression rather than journalism, this interpretation of what people do online is at odds with the varied cultural meanings attached to online content generation and neglects to consider the fluidity of how media content may shift from one place to another and have different cultural and political meanings over time. In other words, previously one might have treated journalism on the basis of the fixed end products it generates. However, today, those end products are dynamic and always unfinished, and so the labels we attach to them, or the work they do, may shift from being either journalism or personal expression. The original guidelines also subsumed all forms of digital content production under the definition of *blogging*, which itself may be a term that inadequately describes what people do when posting content to social-media platforms. Finally, the guidelines may not adequately account for how professional journalists have begun to occupy social-media environments.

Thus, it remains doubtful as to whether the IOC—or anyone for that matter—is in a position to come to terms with the transformation of journalism and social communication that is brought about by social media. Resolving some of these issues requires understanding how society regards blogging, but also needs an understanding of how the spaces within which journalism takes place, along with the communities that produce them, are changing. What occurs within social-media platforms may have any number of meanings. Certainly, with the capacity of Web 2.0 to permit the endless sharing of content, one's blogging feed becomes just a single component in a wider ecosystem of journalistic writing. It is already commonplace for professional journalists to use individuals' tweets from Twitter as a source for reports, a mode of journalism that has caused some people to be very cautious about what they say in such a platform. Furthermore, many media outlets draw on user-generated content in their creation of stories and in the discovery of emerging stories, as the rise of Storify attests. Storify is a social-media platform that allows its users to curate content from a range of sources and pull it into a single stream.

Thus, beyond audience members' undertaking their own journalistic work, one of the distinguishing features of the social-media era of journalism is that audiences play a more active role in constituting the practice than was previously the case. For example, if a Twitter user follows the BBC Twitter account and then re-tweets content published by the BBC, then they become an integral part of the BBC's production line by further disseminating its work to more viewers. In this case, the Twitter user's role is to amplify, rather than to take part in the BBC's investigative process. However, if a BBC journalist is following a Twitter user and something the user tweets becomes a lead for a story and the journalist and the Twitter user engage with each other about the issue, then the user may become interwoven with the journalist's own work process. In this case it is clearer how social media may be more comparable to a journalist's picking up a lead through conversations out in the field than to just being a personal diary. Yet the ambiguity of the example reveals how society and the media are in tension when attempting to wrestle with the value and the meaning of what happens in social-media environments.

The example also calls into question the legitimacy of creating rules to govern online activity—a task that may be akin to having a set of guidelines to underpin interactions in everyday social spaces. Certainly there are social norms and laws that may inform what people do within social worlds, but rarely is there a need for written guidelines to ensure reasonable conduct. Moreover, the practicality of enforcing rules in a fast-moving online world may have little practical value. If athletes are required to seek advice from the IOC on whether or not they can re-tweet something before doing so, then the currency of social media will be lost, since the currency of social media relies on the capacity to communicate the real-time live experience. Again, this raises the question as to whether use of social media can be overseen by guidelines at all, or whether top-heavy guidelines only jeopardize its value. However, as with many guidelines, the breach of rules need not mean that the International Olympic Committee or a National Olympic Committee pursues an athlete for failing to adhere to them in all cases.

In this context, a wiser approach to interpreting the IOC's Blogging Guidelines would be to conclude that they are designed to protect the IOC from systematic attempts to exploit IOC property. On this interpretation, sanctions are likely to be directed only toward those individuals or organizations who have made concerted efforts to use platforms in such a way as to advance their commercial interests. It is more likely to be these large-scale infringements that attract the interest of the IOC, rather than individual athletes undertaking minor actions. Indeed, in some environments, infringements by spectators can be addressed subsequently, rather than needing rules that prevent actions. For example, if an Olympic fan films the television coverage of the Games and then uploads that video to YouTube, then it is possible for YouTube to identify that content by using audio visual content tracking and to either remove it or monetize it with commercials. These identifiers now offers more opportunities for organizations to adopt multiple strategies in response to infringements, rather than simply seek to take legal action.

In the same year that the IOC published its initial blogging guidelines, the British Olympic Association published similar guidelines for its athletes, which stated the following:

Online Diaries (BLOGS)—Athletes may not submit journals or on-line diaries to Web sites during the actual Games Period as this is deemed to be similar to reporting from the Games which is not permissible under the Olympic Charter. Athletes and any other accredited participants are free, of course, to respond to questions from journalists, Web editors, or the public, on any site in an ad hoc fashion. (British Olympic Association 2007)

These 2008 guidelines were met with widespread criticism within the UK, as they were seen as even more restrictive than the IOC's—perhaps even contradicting them. However, the reason for this controversy is only partially explained by focusing on concerns about individual liberties. Rather, the British Olympic Association was especially concerned about how athletes might be exploited in relation to the political issues facing China's first Games, which took place amid international criticism. This generous defense of the BOA guidelines is challenged by those who argue that they were actually designed to limit athlete's freedoms, an ongoing concern about how athletes may operate when accepting their contract to represent Great Britain in the Olympic Games. On this matter, it is important to consider the way in which the IOC steers the Olympic program away from partisan political issues in order to preserve what it considers to be the apolitical space of the Olympic program.

These debates reveal how journalism is expanding practice in a digital era, a phenomenon which is creating new questions about what journalism entails, how it is distinguished, and how it should be regulated. The tensions they provoke are played out at the Olympic Games, where the media are steadily being challenged by new communities of journalism with a range of political interests. A number of questions arise from this tension. For instance, YouTube, Facebook, and similar platforms become the primary environments on which audiences share what is happening in the Olympic city. They are not simply another channel that we watch, but are completely new ecosystems of information. They are also the places where people go online to find content. Alternatively, will these environments be more free and open than the traditional, accredited media spaces? As more and more individuals regard themselves as citizen journalists, might one expect a new kind of media coverage to surround the Games? Are professional journalists under threat by such communities? Will journalism at the Olympics be conveyed via blogs and podcasts rather than by television? The following section considers some of these questions by investigating how the Olympic bodies might monetize the user-generated content of social-media platforms to maintain their position as owners of all media content associated with the Olympic Games.

Monetizing the Olympic Web

A decisive influencer in the ongoing expansion of media artifacts at the Olympics is the IOC's capacity to monetize content within new media environments. Yet, while analysis of the Games often focuses on the sports, the key component of this discussion has more to do with the Olympic experience as a period of seven years leading up to the Games. In short, figuring out how to make the most of the time in between Games provides the most compelling way for the IOC to maintain control of its assets.

Historically, the principal concern in terms of the monetization of Olympic content has been control of the IOC's main identity, the Olympic rings. Games after Games have been replete with stories of breaches of Olympic intellectual property and their perpetrators' being pursued by the IOC, by the Organizing Committee, or by the National Olympic Committee. Examples include retail outlets that have created their own Olympics-inspired window displays have been forced to remove them. Indeed, before the 2012 London Games a butcher in England was required to take down a display of sausages shaped like the Olympic rings, which caused the former Head of the IOC's marketing guidelines, Michael Payne, to say that the rules were "never intended to shut down the flower shop that put its flowers in Olympic rings in the window, or the local butcher who has put out his meat in an Olympic display." "The public do get it," he continued. "They do understand that Coca-Cola has paid, Pepsi hasn't, so Coca-Cola should be entitled to provide the soft drinks, but what's that got to do with a flaming torch baguette in a café?" (Payne is quoted in Peck 2012.)

The incidents mentioned above may illustrate the lengths to which the IOC and Olympic stakeholders will go to curtail trademark breaches, but they also suggest that similar breaches online—even if minor—could be subject to similar legal action. Some recent examples of breaches that fall into the category of ambush marketing include the URL

http://www.london2012festival.com, which, instead of taking browsers to
the London 2012 Festival (the main Games-time Cultural Olympiad brand
for the 2012 London Games) took them to a website selling mobile homes.
And shortly after Rio de Janeiro won the 2016 bid, the URL www.rio2015.
com led to a website with a two-frame layout; in the smaller frame was a
rogue advertising banner, and in the main frame there was an embedded
mirror of the official Rio 2016 website. The website is no longer active.

Whether or not organizing committees or National Olympic Committees pursue all such breaches the Internet presents new cases that test established precedents. In all cases, the key question for the IOC is whether to
protect its interests by pursuing all such breaches or by making examples
of some so as to deter further infringements. Alternatively, it may seek to
ensure that any Olympic content is subject to the interests of the IOC and
for digital technology. Monetization rather than censorship may be the
most effective and efficient way to achieve that. Indeed, such strategies are
already apparent within social-media environments. For instance, the IOC
may seek to ensure that videos on YouTube tagged "Olympic" are preceded
by an Olympic-sponsor video. This may prove to be a more effective way of
monetizing content than having a strict "take down" policy.

The financial model of recently successful Web ventures mostly relies on
two principles. The first has to do with the generation of marketing data,
which may then be sold to third parties to assist with the advertisement
of products or services, either within or outside of cyberspace. The second
(a more recent principle) relies on the free utilization of services by a large
consumer base complemented by a comparatively small number of users
who pay for advanced services. A good example of this "freemium" model
of e-commerce is Google, which utilizes data gathered from its search facility to sell such services as Google Analytics to organizations that are interested in tracking their profile online with a view to improving their impact.
Many other start-up companies, including Flickr, YouTube, and Prezi, have
adopted a similar model. That model functions on a simple logic: If the
product or service is good enough, then its adoption by a mass audience
will give rise to communities of premium users, and those communities
will support the development of the software. Moreover, the open-source
varieties of such software furnish companies with a large community of
developers who will work without pay to improve the application because
they value and care about the product. The monetization of the Olympic

movement's digital assets may, then, draw first on the pre-Games period, during which millions of clicks will occur in search of Olympic content, which could then become an integral part of the IOC's rights package. (In order for this to happen, it may first be necessary for the IOC to gain sponsorship from a large online media provider, such as Google.)

These aspects of the monetization problem offer important insights into the context of the adoption of social media within the Olympics, though it would be more interesting to consider how *open-source* media communities fit within a climate in which social media promote the retention of power and control. There are specific issues that are at stake when considering this subject in the context of the Olympic movement; they have to do with how the Olympic movement expresses its social and humanitarian goals. In some cases, the monetization of digital media is in conflict with the Olympic movement's constitution and the ethos of digital culture. Treating social-media assets as part of the corporate marketing of an Olympic organization is inextricable from this monetization approach. Yet many of the Olympic stakeholders undertake such actions in how they use social media, as the next chapter will discuss in the context of London 2012.

The Need for Open Media

One of the disputes about how to handle social media within the Olympic ecosystem concerns whether or not the drive toward greater openness brings some risks to those who previously had sole control of the communication channels. There are two components to the risk that may be brought by such openness, the first of which concerns risks associated with controlling one's narrative and the second of which has to do with risks to one's economic foundation. The two may be closely related. In the case of the Olympic Games, loss of exclusivity in the communication of content by the rights holders can diminish their desire to invest into spending large amounts of money for such privileges. Alternatively, an inability to control the main campaign messages that are produced alongside the Games may make the Olympic Games less appealing to utilize as a communication channel.

The IOC has undergone a number of transformations to its management of the media that have worked toward mitigating these risks. At this point, it seems more apparent that openness of media reinforces the capacity of a holder of intellectual property to advance its interests, but it is also clear that openness brings new challenges and calls for new ways of working. Consider one simple principle that distinguishes social media from other modes of communication: that of user-generated content. Never before has an individual had such capacity to destabilize the information hierarchy as has been made possible by the generation of user content. Today a dynamic personal website can be more powerful than a static institutional domain in terms of recall by means of a search engine such as Google.

This challenge to institutions is real, but it exists regardless of whether or not an institution makes its communication channels social. Indeed, an organization is more likely to lose its authority over its brand's identity if it resists embracing the sharing economy. Failing to adapt to the manner in which communications occur in public spaces rather than private channels entails a loss of opportunity to assert one's voice. However, the more persuasive argument for opening up media channels to contributions made via social media involves taking into account how Internet users migrate from one platform to another. Thus, if an institution or an organization is not present in a major social-media environment such as Facebook, it will reduce its contact time with its Internet audience, simply because that is where the audience experience online takes place. This is evident from the number of user-generated "groups" dedicated to the Olympics that can be found in such environments.

These aspects of social media reveal why it is risky for an organization to fail to adapt to the changing digital sphere. Moreover, without a strategic approach towards new digital platforms, detachment from one's audience and community, as it develops within such spaces. This is not to underestimate the dramatic implications of such a shift for institutions such as the IOC. Indeed, for any large, transnational organization, developing an adoption strategy that does not jeopardize the effectiveness of existing contractual arrangements is challenging. Yet there are even more reasons why it is important to pursue an experimental approach to using social media. After all, as revenues from new media increase, the Olympic proposition may become less attractive if it fails to come packaged with new-media rights. It may be increasingly difficult for the IOC to retain global sponsors if it fails to stimulate innovation in this area. Indeed, there are good reasons to believe that the IOC understands this, as the Olympic Channel suggests.

Trusting the social-media community is a crucial part of this process, and various Olympic moments have revealed that it is possible to relinquish control of the communication channel without compromising on strategic ambitions. A good example of this was provided by the dress rehearsals of the London 2012 opening ceremony. Such rehearsals run the risk that highly guarded content might be made public if someone shares photos or videos of them on social media. There is very little that the organizers can do to prevent this, and so they must rely on trust between organizer and audience. To encourage such trust, the London Organizing Committee used digital screens within the venue to ask the live audience to "Save the Surprise." They also used the Twitter hashtag #savethesurprise. This dual message led to thousands of tweets in which people shared jokes about the ceremony based not on the actual content but on what people who were not there might imagine it to include. In this case, Twitter was used as a vehicle for entertaining, not informing, and that discourse became part of the Games narrative.

Expanding the Olympic Experience

As was suggested above, the monetization of Olympic assets is likely to have value particularly for the non-sporting aspects of the Games, partly since there is a lot more flexibility in what publishers can do with these aspects of the Games. This includes the cultural activities, the street environments, and the large fan zones, along with what happens in between the Games. Whereas the sports events are relatively fixed in terms of their format and the contractual relationships around their production, everything that happens outside of the venues is much more varied. However, it does require a change in approach to the Games, from thinking of them as just sports events, to thinking of them as cultural festivals. Indeed, there are various indications that suggest sports mega-events are becoming treated in such a manner by host cities and examples of innovative ways in which digital media is maximizing the exploitation of such interests. There are also various layers to this proposition, but let us first consider three types of digital assets: Cultural Olympiads such as the Vancouver Cultural Olympiad Digital Edition CODE), Facebook groups, and Twitter accounts and feeds.

The Cultural Olympiad (CO) is a special case with regard to monetization. Traditionally operating outside of the core brand identity, the CO has a history of struggling with the intellectual-property arrangements required by the IOC. Even ownership of photographs becomes an issue of concern, particularly when they are shared online. There are some obvious examples of how conflicts arise around the branding of culture generally, but what specific concerns may emerge about a digital cultural artifact? A good template for these considerations is found in the Vancouver 2010 CODE project. Internet users—mostly Canada based—were asked to upload a photograph that conveyed their understanding of "Canada" to others. Thousands of photos were uploaded, curated, and then presented and using a website design that placed the images randomly within the CODE website. Once uploaded, these photographs became the property of the organizing committee. This was an unusual transference of intellectual property in a world where Creative Commons licensing has led to the responsible sharing of intellectual property between individuals and organizations. A crucial facet of the project was the role of curation, although *moderation* may be a more appropriate word. After all, Vancouver's Games—much like others—involved considerable protesting by disaffected local communities whose images were absent from this digital celebration of Canada. One reason why an official Olympic digital asset cannot yet fulfill its social role as a space of social media has to do with the compromise brought about by the need for the content to align with a set of Olympic values. In this case, the contractual obligations behind the Cultural Olympiad—which the government of the host city agreed to when signing the contract with the IOC to protect the Olympic brand—restrict the kinds of content that can be elevated. In this moment of Olympic celebration, the host city willingly relinquishes some of its freedom to permit the demonstration of political views, insofar as they run counter to the Olympic values.

A second example is a Facebook group. Facebook is interesting to study in the context of the Olympics, because of the fluidity of how intellectual property is managed. For example, if a Facebook user uploads a video including some content that is owned by someone else, Facebook will endeavor to identify this and prohibit its publication. However, it may be possible create a group or a page that uses the graphical emblems of another organization, even though this may also be a breach of intellectual property. For example, in 2008, when a decision on the site of the 2016 Games was to be made, there were numerous "support" pages for Chicago's bid. One of them was managed by the bid organization, but many others that were not managed by that organization used protected emblems when promoting their page.

The ability to take such action and not suffer some kind of legal response from Olympic organizers may have something to do with Facebook's ambiguous public-private status. After all, only members of Facebook could see the infringed content, and so technically, it occurs within a private space—which is not subject to the same kinds of restrictions. Yet to the extent that a large proportion of the population is now present in Facebook, it may no longer make sense to talk of it as a private location. Indeed, in 2013 Facebook's CEO, Mark Zuckerberg, pronounced privacy "dead." This has particular resonance given the way that Facebook has opened up more of its content to public viewing in recent years, a process that has resulted in considerable controversy as a result of its implications of failing to recognize the privacy of users who had signed up for a private space rather than a public environment.

Monetization in these social-media platforms occurs on numerous levels and in different ways. Through Twitter, an Organizing Committee for the Olympic Games (OCOG) can draw its audience to content that it prescribes. It may even permit sponsors to tweet content on an agreed and automated basis, which provides even further opportunity to amplify the sponsors' brand recall. Facebook's AdSense technology ensures that sponsors obtain benefit, while the content management capabilities of the application could easily permit the sharing of film and image content by sponsors, thus furthering the IOC's reach.

With regard to a digital asset such as CODE, the fact that the technology emerges from the OCOG's new-media infrastructure ensures that intellectual property is retained, but its not being located within a public social-media platform limits the content's reach considerably. Furthermore, these circumstances deem that the content will have nearly no legacy, since it will be tied to the OCOG's website infrastructure, which will also be compromised when the OCOG ceases to exist. Though any specific initiative can be archived in digital form, this alone does not ensure that it has a meaningful legacy once the Games have concluded. And yet the entire conversation about the value of an archive in an era of social media has become highly complicated. After all, one of the most successful social-media platforms is Snapchat, which focuses on delivering content that it will delete after a short period of time. Historically, this has enormous implications since it invites us to consider the impact of a world in which what is shared through communication becomes untraceable in the future. The loss of history that is brought by this structure is significant, but thirty years ago the potential for such dialogues to be history did not exist. In this sense,

Snapchat's approach may resonate more with the concerns that too much of our lives is archived and that there is value in resisting this compulsion. After all, the endpoint of continuous recording of our lives is an inability to reflect on what we are doing—already the amount of content we generate in film and image exceeds our capacity to even look back over the content once it is created. In a world in which content generation is continuous, this loss of reflection is even greater.

What are the enduring features of these case studies that may assist studies of digital culture in the future? Are there trends that indicate emerging ways of thinking about our relationship to communities, brands, and products? Alternatively, are any attempts at analyses likely to be redundant within two or three years, once new platforms have emerged? How will the major social-media platforms look ten years from now? Will Facebook still exist? Will YouTube? Will Twitter? In advance of the 2016 Rio de Janeiro Games, NBC Olympics had already indicated that they would utilize Snapchat as a principal communication medium, where previously they had focused on Twitter. Indeed, many young people already have switched from Facebook to Snapchat and Whatsapp, which allow one-to-one private chat, and the currency of public sharing may be in decline. NBC has had a long history of innovating within social media, and its having paid for the rights to broadcast the Games in the United States until 2032 requires it to stay abreast of new digital trends. In the end, switching strategic ambitions to the particular platform that is in vogue at the time may be the smartest approach to take with social media. Such an approach also has the added value of being seen as innovative.

What seems clear about the future of mega-events such as the Olympic Games is that the venues are just one component in a wider festival experience that is being developed and played out in other spaces, whether these are simply outside the arenas or online in social-media platforms. In each case, the way in which an event is conceptualized and produced must change, taking into account the number of remote participants who may be physically close to but who are outside of the official venue zone. Hosts of future Games will have to find ways to optimize the feeling of participation by both kinds of community. In this regard, the most compelling propositions are focused on creating a synergy between the physical world and the digital experience. Twitter and similar platforms have transformed how people access information or news, but also because they are utilized especially as mobile platforms. By this I do not mean just that they are used on mobile phones; I mean that, whereas being online once meant being indoors, behind a computer, it now means being outdoors, on the move, traversing physical space with a digital interface.

Digital Olympic Volunteers

There is a lot of commonality between the traditional Olympic volunteer ethos and the community of content generators that are found online. Investigating this commonality can be of considerable value. One of the clear manifestations of the symbolic capital generated by the Olympic Movement is found in people's willingness to volunteer to help produce the event. (Summer Games require about 70,000 volunteers.) Indeed, understanding what leads people to spend their time undertaking such work may provide helpful insights into what kinds of things users of social media would be willing to do in support of an event they care about. What if, instead of being developed, managed and owned by the IOC, they were developed, managed, and owned by the volunteers, or by a new community of volunteers?

There are various positive consequences that could result from such a shift in how digital development takes place within the Olympic community, though it is useful first to understand how this principle is integral to the economic proposition of some of the most successful software applications of recent years. Consider Wordpress, a popular tool for publishing on the World Wide Web. Wordpress makes it possible for developers to integrate their own design qualities via software "plug-ins"—short pieces of code that Wordpress can integrate to affect the overall functionality of the core template. This model has allowed Wordpress to become one of the main market leaders in Web publishing. It is not used only by bloggers; even the website of the magazine *Time* is powered by Wordpress.

Increasingly, companies are opening up their software for development by third parties through Software Development Kits (SDKs), and this is seen as a way of brining new innovation to platforms and allowing them to operate in different scenarios. Olympic audiences can be empowered through such change, which allow them to play a more crucial role in shaping the Olympic program.

The Olympic industry is distinguished from other such events by its ideological constitution, the Olympic Charter, which highlights its broad humanitarian aspirations. The mechanisms through which these aspirations may be advanced should be of direct concern to the IOC, particularly in a period of media change. Traditional media are in a time of crisis, brought about in part by the proliferation of user-generated content and the appropriation of journalistic values and methods by citizens. Newspapers have fewer and fewer readers. Television is operated by monopolies. Even the practice of journalism is under threat by a wider proliferation of information by citizens. For this reason, the IOC's relationship with its media stakeholders requires re-definition, so as to find a model that works in this changing climate.

By emphasizing social media and user-generated content, the IOC can re-energize its audiences, which may have become disenchanted with the IOC's overly commercial and proprietary model. Indeed, the activity operating around the IOC's social-media campaigns is indicative of this. However, the broader point about this need for a shift concerns changing patterns of labor in a digital era, which changes what we regard as the institution and its public. In other words, in a world where the majority of the content around an Olympic Games is generated by the public, rather than the event's legal owner, societies are forced to consider the legitimacy of such private ownership. This is why digital technologies should be understood as activist devices: They can change the conditions of how institutions relate to individuals, or how the governors related to the governed. The politics of such activism need not be anarchic or even opposed to dominant forces in society, but they should always be built on the premise that audiences are both producers and consumers of social artifacts, like the Olympic Games. Though the rhetoric of such circumstances has been a driving principle of media production for several years, it has rarely led to transference of power toward the audience.

If the empowerment of stakeholders or the participatory community isn't enough to encourage a change in operations, the ability to develop market share may be sufficient. The concept of market share is not easy to apply to an Olympic Games, though a reasonable set of indicators arises from the concept of television audiences. Again, this concept has a different currency in a digital era in which the units of "clickthroughs" or "shares" may surpass or at least complement page or channel viewing figures. Today, the idea of "age views" is also a growing measure of online impact. Yet when working with an audience as a development community, the opportunity to advance page views grows exponentially. This is partly to do with how development works online. By building a development community around a product, it is possible to transform audiences into a critical mass of brand ambassadors whose labor becomes an integral part of the brand's evolution. Kücklich (2005) describes such a process as "play-bour," arguing that the "relationship between work and play is changing" in such contexts, the effect of which is to re-inscribe "the consumer in to the production process" (Kline et al. 2003, p. 57).

The ability of an independent organization to undertake such a feat without public mobilization is limited because the success of any such enterprise relies, increasingly, on the labor of many hundreds or thousands of participants, which a company cannot expect to staff independently. Another example of the model is Twitter. Since its inception, Twitter has steadily been appropriated by the news media, leading to changes in how they distribute news. The dominance of the mass media within Twitter is also important to acknowledge, not least because it brings into question whether the rise of new media brings about any change within society at all. After all, many of the top Twitter accounts are owned by mass-media organizations. This is true of other social-media platforms too. For example, data from January 2016 shows that the most shared content on Facebook belongs mostly to traditional media outlets such as Fox News, the *New York Times*, NBC, the *Guardian* and the BBC (Corcoran 2016). All of this suggests that social media function as a new medium rather than a new-media outlet. Whereas audiences are used to categories of media formats in the past as being print, television, or radio, individual social-media platforms may have a comparable status as categories, rather than part of some wider category we call social media. Thus, Twitter is better analogized to radio, television, or print media than to some wider category of online media. However, this situation alone does not confirm that no change has occurred, as it is partly indicative of the fact that such companies arrive at new applications with established followings and as reputable mediators of cutting-edge news. In this sense, it is reasonable to expect the media to continue their dominant position.

Another way of assessing media change is to identify how an average Twitter user accesses information and, in this regard, it is clear that the sources of news have become more fragmented in the digital era. Equally, the environments where news is found have shifted and no longer rely on hierarchical rankings. For example, the free Twitter browsing device Tweet-deck allows a user to prescribe what they see by setting up curated Twitter lists, which respond to specific search interests. In other words, what distinguishes the era of social media is that there is no generic way that everyone uses the platform, no single channel that everyone experiences. There is simply a filter bubble of our own making. Such environments effectively blur forms of personal correspondence with news content, when often the former also performs the action of the latter.

Yet it would be naive to imagine that such platforms as Twitter herald a news era of one-to-one or even many-to-many communication. Indeed, the recently introduced "Moments" option within Twitter provides a collaboratively editorialized stream of Twitter content, albeit a stream that is reliant on trending topics within the platform (Caddy 2015). While social-media platforms trade off the idea that it personalizes communications, a considerable amount of Twitter activity is shared automatically. Indeed, such applications as If This Then That (IFTTT) allows uses to push content to Twitter simply by setting up a link from a website's RSS feed to a Twitter account. To many, the content may appear as a personal act involving some manual labor, though often it is not. Equally, social media may easily become antithetical to sociability, via the rise of Twitter spam and Twitter porn, which have been present in Twitter since its inception. In this sense, it is hard not to conclude that one end consequence of the proliferation of social media is a removal of inter-personal connectivity, in exchange for a system of automated sharing devices. One begins to see such nuances within prominent social-media platforms which have gone through a long period of providing a key service, but which then have begun to monetize the platform, as for promoted tweets in Twitter, or Facebook Ads shown in users' news feeds. Indeed, one must always conclude that the business model of social media is first to capture a user's attention, make the use of the platform a daily habit, and then exploit this habitual behavior by sprinkling advertising into the environment. By the point at which this monetization occurs, the users will be so tethered to the platform due to its value in assembling their social network, that the monetized content will be accepted reluctantly, but certainly. For this reason, the long-term goal of social-media platforms is for them to become highly personalized social advertising environments, where the user is so interested in the content that they do not even perceive it as advertising. However, there are also greater aspirations that surround the rhetoric of owners of social media, as might be said of Mark Zuckerberg in his aspiration for the company to bring the Internet to parts of the world where it is not yet available.

Overall, social media have had a dramatic effect on how the Olympic Games organizes its communications and how it mobilizes its community. Social media now are where news is broken, where new audiences are formed, and where new streams of revenue are generated. These three changes have become cornerstones of the Sport 2.0 era, which is demarcated by a shift away from more traditional media formats. The next chapter focuses on the institutions responsible for creating content and consumer communities, before returning to a discussion about how new communities of citizen journalists are taking ownership of the channels of communication around the Olympic Games.

9 The Effect of Social Media on the 2012 London Olympics

As was outlined in chapter 8, Olympic media organizations have been quick to demonstrate their capacity to innovate from one Games to the next, and their use of social media has been no exception to this rule. In this chapter, I examine some of these initiatives, looking at the 2012 London Games as the first, comprehensive social-media Olympics. I consider the range of ways in which social media have been used by examining campaigns by various Olympic stakeholders. Building on the previous chapter, I argue that content generated for social-media platforms reveals important shifts in how journalism takes place and in what audiences identify as journalism content, but that it also blurs the boundaries between journalism and entertainment.

As more media outlets provide increasing amounts of space for the public to comment on journalism, the discussions within these spaces become a more influential part of our media culture, calling into question the role of the initial stories—and, by implication, the primacy of professional journalism—on which they are based. A good example of this is "reply videos" on YouTube, which may often attract more views than the original content to which they are replying. An example that is more closely located to formal journalism is found in the Huffington Post, an online news platform. Through its bloggers, many of whom might be described as public intellectuals rather than journalists, the Huffington Post is able to develop a significantly larger volume of content and to attract contributors who would otherwise publish their ideas elsewhere. This approach to creating a wider range of volunteer contributors was adopted by numerous other media outlets ahead of the 2012 London Games, including the free-newspaper network Metro, which appointed 100 bloggers to produce content for its website. Other media organizations have been quick to exploit the currency of public comments on stories and have even handed over their own domain space for their readers to write and publish their own stories. For instance, the UK's *Guardian* has a vast online presence and recently created professional portal websites that are, effectively, blogging spaces for users.

These changes increase the number of stories a media outlet can produce (and own). They also encourage bloggers to re-publish their content on an established media outlet's domain, where it can control the monetization surrounding it. In this manner, a media outlet also becomes its own social-media platform, as it now hosts the content rather than leaving it on some other independent or isolated website. In short, a media outlet's appropriation of the intellectual labor of bloggers transforms it into an aggregator or a curator, rather than a generator of content. In turn, contributors strike a bargain when entering into this transaction with a media outlet. For instance, where there is some editorial oversight over what gets published, then such an agreement may restrict what an author is able to publish. Alternatively, these agreements also ensure that independence diminishes and the most successful aggregators grow their market share within the attention economy. The tradeoff is that the author's work gains greater visibility from the larger platform. The challenge for the media organization is a possible loss of control over the content associated with its brand, or a more diluted form of journalism.

Social Media as Generators of News

Outside of the new configuration of journalists' roles and relationships, there is also evidence of how content generated for social media exposes gaps within our media culture, particularly around the generation of news. The Olympic Games have demonstrated this at moments when a concern about the professional media has gained attention through its proliferation within social media. For example, on the night of the opening ceremony of the 2012 London Games, the hashtag #NBCfail began to trend on Twitter as a result of criticisms of NBC's live television coverage, a narrative that continued throughout the Olympic and Paralympic Games (Carter 2012). The criticism was focused on the lack of live coverage, the failure to broadcast aspects of the ceremony, and the interruption of content by commercials. In response to this, the British journalist Guy Adams tweeted the email address of the president of NBC Olympics and invited people to send their complaints directly to him. As a result, Adams was temporarily suspended from Twitter for breaching its guidelines, which provoked further debate about whether this person's email was private or not and whether Twitter should be involved in policing such incidents at all. Yet the more interesting point here is that it shows how original news stories are generated within social-media platforms and how a Twitter hashtag can quickly become an influential force within the news agenda. The #NBCfail period is a good indication of how user-generated content can shape the news agenda and of how what happens in social media has come to influence the creation of news. Yet there must be some strategic value in NBC's approach, as the same story was repeated in 2016 at Rio, where it was criticized for the volume of commercials in the Olympic Games Opening Ceremony.

The Olympics on YouTube

The increased centrality of social media to Olympic journalism is also broadly replicated across the campaigns of other Olympic stakeholders, which have expanded their public-relations and communications programs to emphasize social media as vehicles for reaching audiences and customers. This transformation is also further indication of how it is often ambiguous whether media content is news or whether it is commercial advertising. For example, during the run-up to the 2012 London Games, the London Organising Committee of the Olympic and Paralympic Games (LOCOG) used YouTube to publish a range of films that promoted various aspects of the Olympic program across the seven-year Olympiad period. Their 357 films generated more than 12 million video views and 52,000 subscriptions by the end of the Games.[1]

From the content, a number of insights may be derived about the value of such an undertaking. First, there are clear demarcations of priorities across the years 2008–2012 which may suggest a template for Olympic social-media marketing.[2] In 2008 and 2009, videos focused on the handover ceremonies in Beijing and on construction progress within the venues. In 2010, videos focused on engagement campaigns, such as volunteer opportunities (Games Maker program) and the Inspire Mark events program, which was a system by which people could become affiliated with the Olympic program. Also in 2010, the Olympic and Paralympic mascots were given priority. In 2011, most videos focused on explaining key aspects of the Olympic Games and how it would happen, often involving British athletes or celebrities to talk about what the public could expect from the Games. In 2012, the bulk of the YouTube content in the London 2012 Channel—which represented more than 50 percent of the total videos uploaded by LOCOG—consisted of "behind the scenes" content from the ceremonies, stories about the mascots, educational videos explaining how Paralympic sports are played, and, most of all, Torch Relay highlights. The top ten Torch Relay videos in view count are listed here in table 9.1. The most-viewed videos on the London 2012 channel are listed in table 9.2.

Aside from the Organizing Committee, the IOC has also used YouTube extensively since 2006, though it didn't really begin to generate content until around 2008. Its channel is used to deliver Olympic sport content to countries that do not have any broadcaster agreements, but it also includes material that is never broadcast elsewhere. For example, the IOC Congresses, at which cities present their bids, are now broadcast live over the Internet by the IOC, in part because they are seen as niche interest events, unlikely to appeal to a mass domestic audience. Their being broadcast is a powerful act of transparency that makes it easier for researchers and the media to witness what takes place in full and is a good example of how the unlimited server space provided by social-media platforms can ensure that a lot more is made public than ever before. This approach mirrors others that are happening within established political spheres where government hearings and parliament debates are now broadcast as a matter of course in some parts of the world. While these broadcasts do not provide the whole picture, they are far in excess of what was made globally visible before such platforms existed.

As of June 9, 2016, the IOC's YouTube channel has 516,519,478 video views and 1,255,141 subscribers. Its most viewed video, with more than 17 million views, is the full replay of the men's 100-meter sprint final at the 2012 London Games, in which Usain Bolt won a gold medal. Its twelve most viewed videos are listed in table 9.3. However, 2013, when the data were captured, four new entries have made the top three entries; they are listed in table 9.4. The new entries are videos of entire events, whereas previously the channel had shown only highlights. This might explain how these films have risen to the top twelve videos on the IOC's channel two years after the London Games as sports fans watch the entire replay of a special match. These examples are also interesting from the perspective of the broadcast-rights agreements, since they reveal how the principal

Table 9.1

Olympic Torch Relay Day 68 Highlights—London 2012 [The Olympic Flame travels from Harrow to Haringey, visiting Wembley Stadium and Alexandra Palace.]	86,768
Olympic Torch Relay Day 69 Highlights—London 2012 (This is actually day 70, but there is an error in the video title.) [After 69 days travelling the length and breadth of the UK, today the Flame embarked on its final journey towards the Olympic Stadium for tonight's Opening Ceremony.]	74,105
Olympic Torch Relay Day 68 Highlights—London 2012 (This is actually day 69, but there is an error in the video title.) [On the penultimate day of the Torch Relay, the Olympic Flame travels through the heart of London visiting iconic landmarks on its journey from Camden to Westminster.]	50,259
Olympic Torch Relay Week 1 Highlights—London 2012 [Highlights from the first week of the London 2012 Olympic Torch Relay. Watch some of the amazing scenes as the Flame is lit in Greece, arrives at Land's End and starts its 70-day journey around the UK.]	49,020
Olympic Torch Relay Day 67 Highlights—London 2012 [The Olympic Flame is carried by 144 Torchbearers, it travels on the London Underground and visits Kew Gardens on its journey from Kingston upon Thames to Ealing.]	40,159
Olympic Torch Relay Week 9 Highlights—London 2012 [Watch the highlights from the last nine weeks of the London 2012 Olympic Torch Relay. In total 10.2 million people have lined the streets to see the Flame.]	39,453
Olympic Torch Relay Day 1 Highlights—London 2012 [Highlights from Day 1 of the London 2012 Olympic Torch Relay. Olympic Sailor Ben Ainslie begins the Relay from Land's End and adventurer Ben Fogle carries the Torch at the Eden Project in Cornwall.]	37,406
Olympic Torch Relay Day 65 Highlights—London 2012 [The Olympic Flame travels on the London Eye and passes through the London Boroughs of Redbridge, Barking and Dagenham.]	34,346
Paralympic Torch Relay—London 2012 [Flames will be lit in all four nations of the UK for the London 2012 Paralympic Torch Relay. Dame Tanni Grey-Thompson talks through the details.]	31,384
Olympic Torch Relay Week 8 Highlights—London 2012 [Watch the action from the last 8 weeks of the London 2012 Olympic Torch Relay. Follow the Flame on its journey around the UK.]	24,458

Table 9.2
Top ten YouTube videos published by LOCOG, by view count.

Video title	Duration	View count as of February 24, 2016
Let the Games begin—London 2012	0:22	1,593,514
Mario and Sonic at the London 2012 Olympic Games (Wii)	2:37	1,347,561
Spurs players try penalties blindfolded—London 2012	1:33	946,738
London 2012 Mascots Film 1—*Out of a Rainbow*	3:43	819,990
Rock the Games, live on YouTube—London 2012	0:53	469,169
The Official London 2012 Results app—free to download now!	1:39	418,086
London 2012 Mascots Film 2—*Adventures on a Rainbow*	4:55	415,620
London 2012 Mascots Film 3—*Rainbow Rescue*	5:00	394,926
Title video for Olympic Games handover show—London 2012	1:29	374,482
London 2012 Mascot Dance	0:52	341,620

arrangements are simply the exclusive broadcasting within a territory over a given period—notably the live experience. Here again, the importance of being the most prominent live distributor is made apparent, a growing currency in the era of social media. The competition to claim the largest live audience is a key facet of what is driving strategic ambitions in social media today, and its relevance becomes even greater in a world where we are constantly sharing information and where there is no time to spend catching up with the past.

By sharing these full-match archive films, the IOC has found additional ways to monetize the archive footage from the Games on its YouTube channel after Games are over, and this may create a new revenue stream and a new audience. The case also allows us to consider further what it is that rights holders pay for when bidding for the right to broadcast the games. After all, if the Games are located on YouTube after they have finished, then there may be less incentive to pay for that privilege. Thus, these full-length videos reveal that what matters most to broadcasters is ownership of the real-time live experience.

Table 9.3

Top YouTube video view count on the IOC's Channel.

Video clip with ranking on February 15, 2013	No. of views, February 15, 2013	No. of views, January 7, 2015	Ranking on January 7, 2015	No. of views, February 23, 2016	Ranking on February 23, 2016
1. Athletics Men's 100-meter Final Full Replay	5,563,689	12,230,979	1	16,185,018	1
2. Closing ceremony—Spice Girls	5,154,369	8,519,716	2	10,595,427	2
3. Athletics Men's 200-meter Final Full Replay	2,939,089	5,790,647	5	7,363,451	6
4. Rowan Atkinson Sequence— opening ceremony	2,643,364	6,620,704	4	9,642,314	4
5. The Queen Sequence— opening ceremony	2,143,649	5,545,214	6	8,371,936	5
6. Closing Ceremony—Queen featuring Jesse J	2,032,348	3,556,679	8	4,480,563	9
7. Usain Bolt World Record— Men's 100-meter	1,978,177	3,256,218	9	3,679,729	11
8. Closing Ceremony—Emeli Sande performance	1,874,432	2,754,528	12	3,008,208	15
9. Athletics Men's 4 × 100-meter Relay Final	1,507,421	4,516,596	7	6,793,140	7
10. Closing Ceremony—Take That	1,311,433	2,168,098	16	2,500,196	22
11. Closing Ceremony—John Lennon, "Imagine"	1,118,282	1,867,137	20	2,361,926	24
12. Closing Ceremony— Wonderwall	1,082,778	1,634,888	22	1,877,839	33

Table 9.4

Video clip with ranking on February 23, 2016	No. of views, February 15, 2013	No. of views, February 23, 2016
3. USA vs. Nigeria—USA Break Olympic Points Record—Men's Basketball Group A from London 2012	6,723,417	10,331,798
8. Michael Phelps Wins Gold—Men's 100-meter Butterfly	Not captured	4,840,505
10. Women's Beam Final from London 2012	2,992,806	4,251,087
11. USA vs AUS Men's Basketball Quarterfinal (USA Broadcast)	2,868,665	4,474,493

There are further insights one can derive by examining the IOC's YouTube viewing statistics. For example, the data reinforce the importance of the ceremonies in the Olympic program, which account for some of the most viewed videos. However, the data also reveal a greater prominence for the closing ceremony compared with the opening ceremony, which challenges the commonly held view that the opening ceremony is of more interest to audiences. Each of these clips is far from being the biggest video on YouTube of 2012, but it is interesting that sections from the 2012 London Games dominate the IOC's present catalog of videos. It is also interesting to discover that these videos are of non-sporting aspects of the program, notably the comedic and musical sections of the show.[3] One might conclude from these films that an organizing committee should invest at least as much in its closing ceremony as it does in its opening ceremony; however, that does not tend to happen, since the largest live audience for an Olympic event is that for the opening ceremony.

How the London Organizing Committee Used Social Media

In August 2012, one day after the 2012 London Games had concluded, LOCOG's Head of New Media, Alex Balfour, shared data about the organizing committee's social-media achievements. Two of their claims were that 46,000 people had subscribed to LOCOG information when London won the bid and that there were 500 million in the email database by the time of the Games. There were 109 million unique visitors to the London2012.com website, 431 million visits, and 4.73 billion page views. The mobile app designed for mobile devices was downloaded 15 million times, achieving number-one ranking in seven countries.[4] There were 4.7 million social followers on the four main platforms used by LOCOG (Foursquare, Google+, Twitter, and Facebook), Twitter and Facebook contributing more than 3 million toward that total. In six years, LOCOG's new-media team commissioned and managed, entirely or in part, 77 digital products, sites, or services, including the following:

London2012.com
Get Set (education site)
School Leavers site
Pre-Games training camp venue site
Velodream competition
Gamesmaker (volunteer) site and sign-up platform (with Atos)
Torch Relay site and map
Torch Relay nominations platform
London Prepares site
Memorabilia auction site
Online shop (with ecommera)
Mobile site
Recruitment sites (ODA and LOCOG)
Local Leaders site
Event database and front end site
Mascots site
Learning Legacy site
Festival 2012 site
#1yeartogo Twitter visualization
Young Gamesmaker site
Ticket sign-up
Ticketing site (front end html only)
Open weekend site
Travel advice for businesses
Twitter, Facebook, YouTube, Google Plus, and Flickr accounts
"Join in" App on three platforms

Balfour notes that "66% of Web traffic to the principal site came from search [engines] during the Games," that there were "10 Google doodles in 16 days," and that "5% of traffic came from Google 'knowledge pan-els' and the doodle." Furthermore, Balfour notes that Facebook was the "top referring traffic source after search" and that uptake of the app was generated by arrangement with all app stores. The mobile app displayed more than 15,000 events using augmented reality and computer-generated venue tours, along with "social check ins, top tips, venue histories, weather forecasts," with 10 million unique visitors looking at the Torch relay con-tent, 25 percent of whom saw the map. The mobile app also integrated Twitter details for many of the torchbearers. Key data in terms of the global reach included visits from "people from 201 territories" people from "155 countries used our apps," and "40'% of all Britons visited the website as did 29% of all online New Zealanders, 19% of all online Canadians [and] 12% of all online Americans" (ibid).

Within Facebook, Balfour notes, there were 1.86 million "likes" across eight accounts, and that content reached 49 million Facebook users, of whom 43 percent were under the age of 24. On Twitter, there were 1.9 mil-lion followers across 48 official Twitter accounts (36 sport accounts with live feeds, two mascot accounts, and six event cameras). On Google+, there were 818,000 followers of a dedicated "ceremonies explorer," bringing addi-tional content to viewers and a backstage live blog during the opening and closing ceremonies. There were 60,000 Foursquare followers, with people from 120 countries checking in to "special Olympic and Torch relay badges on Foursquare" and QR codes on all spectator publications. The Mascots Games site received 4 million visits, and users created 150,000 mascots. One of the most successful campaigns for the London 2012 Festival was associated with the artist Martin Creed's work *All the Bells*, in which a vir-tual bell was rung by 66,000 app users on the morning of July 27. This event complemented the offline experience, whereby all citizens of the United Kingdom were asked to find a bell and ring it for three minutes all at the same time to welcome in the Games.

Data of this kind can be difficult to assess from one Games to the next, not least because the benchmark comparisons are always changing. This is true both in terms of the absolutely comparisons that one may make within a single platform, such as Facebook, and across other platforms, where approaches to assessing engagement may be vastly different in terms of the methodology to evaluate achievements. Indeed, from one Games to the next, organizing committees can often claim that they have done more than the previous Games, but this doesn't always factor in simple variables, such as what is the total population of a platform's user community. During 2012, Facebook's monthly number of users was 1.06 billion (Statistica 2016); by 2016 it was 1.65 billion. Often comparison statistics from Olympic Games Organizing Committees do not such simple variables take into account when presenting their data on users of social media.

How the London 2012 Stakeholders Used Social Media

Beyond the LOCOG's social-media activity, a range of digital content was generated by the regional Olympic stakeholders, which were delivering programs of activity across the Olympiad. For example, the Inspired by 2012 program created thousands of events, each of which had a website and a social-media presence, and had a major contribution to the activity that took place online around the 2012 London Games, not least because there were often large institutions (museums, galleries, or local authorities) associated with these programs. Furthermore, even the activity of official London 2012 programs, such as the London 2012 Festival, is not adequately captured within Balfour's data summary, since it also had additional domain spaces and digital assets that sat outside of the principal London 2012 online domains.

Indeed, this highlights one of the interesting tensions within social media: tension between individuals and institutional accounts. Although resources are often used to support institutional accounts, it is apparent that users value direct contact and personalized conversations in social media. On this basis, it is advisable that institutions invest more resources into assisting individuals in promoting content through their identities and, in the case of the 2012 London Games, it was apparent that this can work, if done well. For instance, the Director of the London 2012 Festival, Ruth Mackenzie, began to create a Twitter profile (@ruthmackenzie) in 2009 and, after being appointed to LOCOG, became a very active Twitter user whose productivity and number of followers doubled between March 2012 and July 2012. In this case, Mackenzie's direct and personal interactions were an effective way of engaging followers, who then also shared content generated through her account. The chair of LOCOG, Sebastian Coe (@sebcoe on Twitter), who was the primary figurehead for the 2012 London Games, had considerably more followers (39,729). Yet Coe's content was considerably less personalized and active. In fact, the @sebcoe account shows only thirteen tweets over the period March–July, for which @ruthmackenzie shows 1,400. Furthermore, the number of followers of @sebcoe increased by only 30 percent over the same period, while followers of @ruthmackenzie increased by 100 percent. And the @sebcoe Twitter account showed only minimal interactions, suggesting that it was run principally by a marketing team rather than by Coe himself. This kind of approach to using social media is rarely rewarded online. In this sense, the @sebcoe account—the most influential LOCOG Twitter account after @London2012—did not really capitalize on its potential to engage followers during the Games.

Another learning point for event managers may thus be that, beyond the brand, people want to follow other people, not just institutional accounts. Though it is possible that the growth of Mackenzie's account relative to Coe's is indicative of the relative ease with which a Twitter user may climb from having a few hundred to a few thousand followers, compared with the challenging of growing an account from thousands to tens of thousands, it is also probable that investment in Coe's account would have generated considerably more traffic and engagement. After all, this is what happened with the other core LOCOG accounts. Even if only 20 percent of Coe's followers were engaged by his content, that would have been more than Mackenzie's followers. On this basis, it is hard to deny that LOCOG lost an opportunity to communicate key stories from the Games through more personal channels.

Beyond LOCOG, a number of other institutions invested extensively into using social media to promote their affiliation with the 2012 London Games, and these organizations can be grouped under the banner of the Olympic family. These included National Olympic Committees, sponsors, and media partners. A number of these partners—both international and national—launched social-media campaigns to drive interest and activity during the Games. One of the platforms used most by each of these groupings was Twitter, which was quickly becoming an essential medium for journalism. Accredited media created some of the most prominent Twitter accounts, an indication of the developing relationships between traditional journalism and social-media platforms. The implications of this relationship were discussed by Ovide (2012) in advance of the 2012 Games:

It is unclear if the interests of Twitter and NBC will always align during the Games. NBC plans to contain highlight clips and other Olympics video on its own websites and mobile applications, which won't necessarily appear on other digital services. Twitter also may tread on NBC's turf by essentially programming the Olympics on its events page. Twitter and NBC said they aren't in competition. NBC's Olympics president, Gary Zenkel, said Olympics coverage anywhere helps NBC by stoking viewer interest (online).

Indeed, NBC had been utilizing Twitter for years. For instance, on the lead up to the 2008 Beijing Games, it created a Twitter list of Olympian accounts, which became Twitter's most followed list. Twitter lists are simply ways to create a single feed composed of the Twitter content from a designated number of accounts and it can be useful to help filter content for users. As a dedicated Olympics account, NBC's Olympic account also dwarfs all other media accounts in terms of followers, having more than 400,000 by the end of the 2012 London Games (which grew from 182,781 just 25 days before the Games began. By comparison, the @London2012 account had 1.6 million followers by the end of the Games (from 718,454 on July 2, 2012) and @sebcoe had 73,000 (39,729 on July 2, 2013).

After the Games, NBCUniversal assembled findings from an extensive audience research program, which revealed, among other things, that roughly "217 million people in the United States watched the London Games, making it the most watched television event in history" (Chozick 2012).[5] NBCUniversal's success in these Games—whereas it had suffering financial losses after the 2010 Vancouver Games—was attributed in part to its strategic social-media partnerships with Twitter, Facebook, Storify, Adobe, and Panasonic. The evidence also indicated that, far from compromising television viewing, the online content was "enhancing the TV experience" (Lazarus, cited in Chozick 2012), which led to NBCUniversal's decision to stream the closing ceremony live.

Within the United Kingdom, one of the most followed Twitter accounts during the 2012 London Games belonged to the UK's official Olympic broadcaster, the BBC. Its @BBC2012 account, along with the Twitter account of the official Paralympic broadcaster, Channel 4 (@C4Paralympics), grew rapidly during the pre-Games period, the latter even surpassing @BBC2012 in followers. Ordinarily the Olympic program would have attracted a larger audience, but on Twitter more people were following the Paralympic broadcaster. This may have been helped by *Meet the Superhumans*, an extraordinarily successful advertising campaign film from the British Channel 4 that captured people's imaginations (4Creative 2012) and the growing sense in which something different was happening in the 2012 Games. This sense of disability becoming seen differently was enhanced by the successful campaign of Oscar Pistorius to compete in the Olympic Games, not just the Paralympic Games.

To date, the two versions of the video on Channel 4's Paralympic Channel—one annotated, the other not—have just over 2 million views on YouTube (1,335,409 annotated, 963,281 not annotated). The same film was located across all Channel 4 programs and advertisements and it won a Grand Prix award at the Cannes Lions International Festival of Creativity. It is worth noting that the view count of this film far outreaches any other films on the Channel 4 Paralympic YouTube channel, the next most viewed video coming in at just 105,348 views and the one after that attracting just 37,803 (as of January 8, 2015). The numbers of pre-Games and post-Games Twitter followers of the dedicated Games accounts of the two UK broadcasters are shown in table 9.5.[6]

The BBC closed its account on September 28, 2012. As of this writing, the Channel 4 Paralympics account continues to post content during preparations for the 2016 Rio de Janeiro Paralympic Games.

Among the most followed other accounts that were set up especially for the 2012 program were that of the 2012 confectionary provider, @CadburyUK (41,563 followers as of July 11, 2012); that of the drinks supplier, @COKEZONE (15,839 followers as of the same date); and those of other media partners, such as @NBCOlympics (182,781 followers as of the same date). Although the number of followers is only one indicator of reach and significance, the numbers provide some indication of relative growth over the Games period and are ordered in table 9.6 by absolute numbers based on the final capture date.

Table 9.5
Twitter follower count of the UK Paralympic and Olympic Broadcasters.

	July 29, 2012	September 10, 2012	January 8, 2015
@C4Paralympics	45,233	120,723	103,378
@BBC2012	31,992	92,334	116,729

Table 9.6
Twitter Followers for Sponsors of 2012 London Games.

Organization	July 29, 2012	September 10, 2012	Growth
@British_Airways	188,950	198,540	5%
@AdidasUK	90,325	124,325	38%
@CadburyUK	48,716	95,828	96%
@SamsungMobileUK	25,073	26,726	7%
@COKEZONE	7,758	21,457	77%
@PanasonicUK	7,724	11,031	43%
@BTLondonLive	3,432	5,726	67%
@EDFEnergy	4,224	5,161	22%
@thankyoumum (P&G)	3,511	4,324	23%
@BPLondon2012	773	1,071	39%
@Omega2012Clock	347	360	4%

Assessing Public Sentiment through Social Media

Beyond managing their own digital assets, sponsors also built novel campaigns around the Games. One of these was the *Energy of the Nation* campaign of the energy company EDF, which created interactive visualizations of Twitter sentiment based on data from around the UK.

Coincidentally, an artist-led Cultural Olympiad project called Emoto—with which I was involved—undertook a similar task, but focused on the global Twitter community rather than the UK. Among their data and reports by others—such as Radian6, the social-media monitoring platform that was used by LOCOG—one clear influencer of peaks is apparent, namely the use of celebrities as vehicles to drive attention. For instance, in the British Airways social-media campaign, the highest peak was achieved by a tweet from Louis Tomlinson of the pop group One Direction thanking "BA for looking after him." At its peak, Radian6 registered 17,496 posts for this day, but the re-tweets continued for several days afterwards. In total, this one message was re-tweeted 28,099 times. Since Louis Tomlinson has more than 5.5 million followers, this clearly indicates that cooperating with such individuals is a sound strategy to drive interest and engagement. However, what is most interesting about the case is that it was not an athlete that had

Table 9.7

Sentiments of UK-based Twitter users toward content of 2012 London Games (source: EDF Energy of the Nation Report, 2012).

Date of games	Positive sentiment	Sentiment title	Related event
2012.07.25	64%	Most positive moment of the day	68% Team GB women's football team beat New Zealand
2012.07.26	69%		
2012.07.27	64%	Highest Number of tweets	47168 tweets per hour during opening ceremony
2012.07.28	72%		
2012.07.29	71%	Most positive moment of the day	79% Lizzie Armistead wins Team GB's first medal in Women's Road Race
2012.07.30	68%		
2012.07.31	64%		
2012.08.01	76%	Most positive moment of the day	76% Team GB claimed two gold medals in the women's pairs rowing and cycling time trial
2012.08.02	73%	Biggest dip of the Games	minus 15% When victoria Pendleton and jess Varnish were disqualified form the women's tea sprint, the nation's positive dipped from 86% to 71%
2012.08.03	76%		
2012.08.04	76%	The Most Positive event	90% Jessica Ennis won Gold in the Heptathlon
2012.08.05	76%	Most positive moment of the day	82% Andy Murray won gold
2012.08.06	74%		
2012.08.07	75%	Most positive moment of the day	82% Brownlee brothers won gold and bronze in the triathlon
2012.08.08	73%		
2012.08.09	76%	Most positive moment of the day	76% Usain Bolt won gold in 200-meter
2012.08.10	73%		
2012.08.11	73%	Most positive moment of the day	76% Mo Farah wins gold in the 5000 meters
2012.08.12	68%	2nd Highest number of tweets	46080 tweets per hour seen during the Games at the closing ceremony

this impact. This phenomenon reiterates the finding from the IOC YouTube channel regarding the role of non-sport content.

From these examples, one may conclude that the social-media ecosystem of an Olympic Games is as much driven by its athletes as it is by artists, perhaps even more the latter, since they are often more internationally renowned than athletes, who are typically more nationally known. In other words, if you want your Olympic social-media campaign to flourish, it is advisable to work with international artists—celebrities principally—by involving them in creating music, making visual art work, or staging performances. The BBC adopted this principle and partnered with Arts Council England—the UK's national funder for art—to create an innovative multiplatform broadcast service for the Games called The Space (@TheSpaceArts), which allowed people to watch live broadcasts of cultural content either online or on a new, dedicated channel on TV Freeview. This may be one of the most remarkable innovation of any cultural program at the Olympics, since often the Olympic cultural events remain only locally experienced, enjoyed only by live audiences. The creation of a dedicated Olympic culture television channel is unprecedented in Olympic history. After the Games were over, The Space remained a place for commissioning new art work, though the platform no longer references its being created as part of the Olympic buildup.[7]

The International Olympic Committee was also proactive in its use of social media during the 2012 London Games, but principally as an aggregator of content. One of the key innovations for the 2012 London Games was http://hub.olympic.org, which brought together athletes' content from Facebook and Twitter, the two principal destinations for the IOC's social-media work.

Facebook and the 2012 London Games

Though YouTube and Twitter are just two indicators of what took place within social media around the 2012 London Games, they are powerful measures of the scale of the activity. However, it would be remiss to exclude other platforms (notably Facebook and Instagram) and the way in which other mobile platforms (such as the London 2012 mobile app) were utilized by the organizers. While the statistics about audience use within these environments is captured in LOCOG's data summary, the breadth of activity across all Olympic stakeholders is far more extensive—and far more challenging to capture in its entirety.

One absolute measure of impact that can be used as comparative data to understand which sports attract the highest follows is the number of "likes" associated with individual athletes. A "like' in Facebook denotes when a Facebook user has accessed an athlete's page and expressed an interested in it by clicking the "like" icon within the page. After the Games, Facebook released some highlight statistics pertinent to the growth in athlete followers during the Games which show that. (See table 9.8.) What is interesting about this data is less the absolute growth in figures than how Facebook articulates what that growth indicates. Thus, Frank and Williams (2012) describe how "local favorites [Jessica Ennis and Tom Daley] became world-renowned athletes throughout the Olympics with their spirit and perseverance" and how some "relatively unknown" athletes "became stars with their wins"—notably Jordyn Wieber (USA), who "won over the hearts of more than 288k," Marcel Nguyen (Germany), and Gabby Douglas (USA). Douglas "wins the 'Rising Star' award" because her number of Facebook "likes" increased from 14,000 to more than 600,000 by the end of the Games. It is in such matters that we see how social-media organizations become more than simply platforms on which users can post content. Rather, they become narrators of what matters sociologically and historically, drawing on their data to support their claims. In this respect, they become storytellers, or journalists.

Table 9.8
Facebook likes of athletes during 2012 London Games (source: Frank and Williams 2012).

	Country represented	Likes during Games (approximate)	Total likes at end of Games
Usain Bolt	Jamaica	1,000,000	8,100,000
Michael Phelps	USA	850,000	6,300,000
Tom Daley	GB	880,000	1,000,000
Jessica Ennis	GB	692,300	801,000
Gabby Douglas	USA	591,000	600,000
Jordyn Wieber	USA	288,000	324,000
Marcel Nguyen	Germany	193,000	200,000

The Rise of Instagram

Around the time of the London Games, there was also a lot of discussion about the use of the mobile photo-sharing platform Instagram. The technology adviser website Mashable reported the following:

More than 650,000 photos have been posted with the #olympics hashtag, while more than 263,000 have been uploaded with the #london2012 hashtag. Even more impressive, a whopping 27,000 photos have been shared with the hashtag, #michael-phelps. Where are people sharing from? Olympic Stadium is the most 'Grammed venue, with more than 7,600 photos posted. Next comes Wembley Stadium; users have posted more than 3,500 photos from the Games' main soccer spot. The Olympic Park Basketball Arena takes bronze, with over 2,300 shares. (Laird 2012)

The success of Instagram may in part be linked to its having been acquired by Facebook in April 2012, ensuring that it became the preferred photo-sharing platform for fans during the Games. This relationship also ensured that Olympic stakeholders prioritized Instagram as the photo sharing platform. This was a departure from the Vancouver Games, during which the IOC worked with Yahoo's Flickr platform to showcase still photography in what was an unprecedented partnership. The IOC's recently appointed Head of Social Media, Alex Huot, had just created accounts Flickr, Facebook, Twitter, and other social-media sites, including the creation of a special IOC Flickr pool where fans could post their photographs. Thus, Flickr may be considered as the first content-sharing environment developed by the IOC and its existence signaled a future in which large new-media organizations such as Google and Yahoo would become close Olympic partners.

The Vancouver Games were also the first Games during which the IOC encouraged fans to post images of Olympic sports on the Internet without requiring permission. At previous Games, such activity led to cease-and-desist letters being sent to some photographers who the IOC felt were infringing its intellectual property. Since Vancouver, this would no longer occur, except in cases where the content was explicitly commercial. Though this was broadly welcomed by bloggers, the Vancouver Games may have also signaled the beginning of the end for the professional Olympic photojournalist, as mobile camera equipment began to allow the creation of print-quality photographs. A case in point is the front page of the *New York Times* of March 31, 2013, which included a photograph first published on Instagram. Although a professional photographer had shot the image, the act of integrating Instagram into the *Times* was a sign of the changing circumstances of professional photography. The photographer's reflections on this are revealing:

That was me in the locker room bathroom shooting portraits of the New York Yankees players with my iPhone. This was not my choice, I wasn't given the option of studio or bathroom stall and decided on the latter. I joined the chain of photographers at 6am in the confines of the New York Yankees Spring Training facility in Tampa, and took what space I could get and worked with it. Below is a set of images shot with my iPhone and processed through Instagram. (Nick Laham, quoted in Shapiro 2013)

The 2012 London Games became a benchmark for how future Games use social media within their program. It exemplified a series of principles that indicate how communication with audiences is changing in a digital world. The lessons from the 2012 London Games are to not underestimate the growth of activity in social media, but also to recognize the need to understand how news is syndicated and how content genres are blurred in an era of social media. Social media enable the Olympic industry to more properly occupy its position as a movement—an entity that mobilizes people across a range of interests and concerns that are of social value. The next chapter focuses on this theme in more detail as it examines what happens at the fringe of the Olympic Games online, as citizens become journalists.

10   Citizen Journalism and Mobile Media

The rise of social media has brought new challenges and opportunities for how the Olympic Games are staged and mediated. Furthermore, it has increased the range of ways in which spectators, fans, and activists engage with the Games, either through public dialogue or simply by enjoying the sports. Social media have also reinforced the dominance of television rights holders, with expected Olympic television viewer time increasing when combined with consumption of social media. In this sense, the growth of social media and (more broadly) the sharing of user-generated content can hardly be seen as obstacles to the Olympic industry's aspirations. However, social media are also changing the kinds of commercial stakeholders that operate around the Olympic movement, and by implication this is altering how the Games are mediated. No longer do Olympic fans simply just watch television; they can now contribute to producing Olympic media content by using their own mobile devices to take part in global conversations about the Games. These conversations become the long tail of the Olympic media infrastructure, spreading out all across the Internet.

Of course, not everyone sharing content seeks to promote the Olympic values, and not everyone desires to celebrate the Games. Some people use social media in attempts to resist the Olympic narrative. This chapter considers the development of these new configurations of media content generation around the Olympics, outlining how new media practices are re-negotiating the otherwise tightly controlled media environment of the Games. Put simply, whereas the non-accredited journalists represent a new kind of professional journalist within the Olympic city, *citizen journalism* reveals the emergence of a new kind of amateur. In so doing, it also describes how media consumption and production can act as a way of "regenerating community" (Jankowski 2006, p. 55) born out of the integration of virtual and physical-world exchanges and a sense of common purpose. While it might be assumed that increased virtual environments would mean that the physical needs of media communities diminish, examples from recent Games suggest the contrary. In considering these changes, this chapter focuses first on the alternative media practices that operated around the 2010 and 2012 Games before discussing their implications for the Olympic media more widely.

My use of the concept *citizen journalism* here intends to demarcate a politically progressive community, which is inherently resistive to the media status quo. In support of this characterization, Kim and Lowrey (2014) speak of the trend of citizen journalism agency to promote a kind of political agency that is grounded in desires for reform and a sense of dissatisfaction for traditional media. In this sense, I also distinguish between these forms of journalism and those that might be found at the Olympic non-accredited media centers, which serve simply to expand the ways in which the Olympic narrative is celebrated. Understood in this way, the category of alternative media—as a distinct form of journalism—may be jeopardized by the rise of social media, if only because social media can crowd out the distinct space that was previously occupied by alternative media. In other words, if there are increasing amounts of Web traffic within the walled gardens of social-media platforms, it becomes harder for alternative media activists to operate outside of these structures, which are also the same structures that they may seek to critique. Yet along with the rise of the Internet and the new forms of democratized self-publishing and self-organizing opportunities it has created, there has also been a professionalization of the alternative media. The models of the Huffington Post and perhaps The Conversation (theconversation.com), along with wider debates about media moguls who many believe have compromised the freedom of the press, have brought new investment into a new kind of alternative media. These trajectories should not be treated as separate from what is happening online in less organized but more citizen-led journalist practice, in which the political orientation of these alternative media is even more diverse.

It is valuable to examine how the use of social media as a form of citizen journalism operates in the context of the Olympic Games, not least because the Olympic family espouses values that identify it as a social movement and not just a sports event. In this respect, there should be common ground between the Games and the media change that happens around social media. Lenskyj (2006) notes that the Games sometimes prompt the creation of temporary oppositional media communities, or to the development of specific campaigns by established media watchdog or independent-media organizations. However, disrupting one of the most tightly orchestrated media events of all time is no small task. Yet Lenskyj (ibid.) also notes that alternative media may be well placed to create a significant impact on the public discourse that surrounds a games, comparable to that achieved by the mass media, particularly within digital space. In this sense, the Internet can level the playing field of political influence afforded to the media.

This chapter focuses on the use of social media by individuals or institutions who operate at the periphery of the Olympic industry. Such individuals or institutions may not be completely outside of the Olympic industry and may even be criticized by their peers for being Olympic insiders when in fact they too seek to question the industry. Before proceeding, I should address the problem of describing these resistant media practices as *alternative*, since that term already relegates them to some marginal position. In my view, this is mistaken since it is more meaningful to think of the work of such communities as complementary to the Olympic accredited media.[1] As I suggested earlier, without these two facets of media production around an Olympic Games, the Olympic industry cannot assert itself as a social movement.

A more appropriate adjective to describe the alternative media might be *disruptive*, since these media practices seek partly to challenge the continuity of the singular messages generated by the celebrating community. Good examples of disruptive media practices are evident in the work of artists who are at the periphery of the Olympic program. In the United Kingdom, leading up to the 2012 London Games, this included work produced through the national arts body called Arts Council England. In this case, while ACE invested public funds into projects related to the 2012 London Games—and while some artists would claim that this reason alone inhibited its disruptive potential—it cannot easily be argued that the artists in receipt of such funding were ideologically or creatively constrained. This is not to say that the ACE program was explicitly critical of the Olympics, or that artists affiliated with it aimed to diminish the celebration of the Games. Indeed, one may observe a certain kind of disciplining effect of the implied relationship that occurs around the Olympic bubble, whereby any major institution finds itself anxious about criticizing the Games for fear of jeopardizing certain political relationships.

However, there are examples of creative work that have criticized the Olympic program, and these instances demonstrate the alternative role adopted by disruptive media. While one might dispute whether art is possible to subsume into a broader category of alternative media, the key point may be that art, like other manifestations of ideas, operates within the media sphere and that works of art become media artifacts. On this basis, it is not meaningful to speak of artwork that exists outside of this networked creativity, and, increasingly, the most compelling media products begin to resemble art work, as might be said of the "Meet the Superhumans" commercial that Channel 4 ran during preparations for the 2012 London Games. Thus, in some cases, these artifacts of alternative media production intersect with major Olympic stakeholders, but often do so in ways that are unrelated to the promotion of sports or other Olympic priorities, as defined by the IOC or by host organizations. These instances alone may count as alternative Olympic narratives, but I am especially interested here in those artifacts that are explicitly, politically opposed to the Games. Collectively, I want to characterize such activities as instances of *ambush media* practice, since they function largely to steer attention to aspects of the Olympic program other than sport and often do so by piggybacking on the visibility of the Games. They are akin to ambush marketing practices, but instead aim to occupy the space enjoyed by traditional media, than just to amplify a particular message or campaign.

Even though alternative media locate their products within mainstream social-media platforms, there is still value in considering what they do as being different from what the mass media do. While the media generally should be seen as agents of community making and remaking (Jankowski 2006), such aspirations are more explicitly prioritized in forms of citizen media and citizen journalism. The roots of citizen journalism are found in the related terminology of alternative media, community media, and grassroots journalism, and these forms of media production have flourished alongside the rise of social media (Allan and Thorsen 2009) and this is largely because

Furthermore, the democratization of professional media technologies made possible by the proliferation of mobile digital devices, increases in broadband speeds, 3G and 4G satellite signals, free apps, and broadcast-quality consumer technology, has allowed citizens to act as organized or accidental journalists, thus contributing to the wider practice of producing news artifacts—text, video, and image. In this respect, the rise of citizen journalism completes the circle of what Beckett (2008) describes as "networked journalism," occupying a formerly awkward place in the delivery of organized journalism through chat shows, talkback radio, and comment. Instead, citizen journalism becomes part of professional journalism, but also carves out its own space, acting as a form of "disruptive journalism" (Beckett and Ball 2012) and as an alternate social space in which news production takes place. In some cases, it espouses the values and ethical principles of formal, professional journalism. In others, it is much more free from control, order, and context—a form of "wild type" journalism, perhaps. The first examples of the development of these new journalist populations occurred in conjunction with the 2010 Vancouver Games.

Vancouver 2010: The Fifth Estate at the Olympics

The Vancouver Organizing Committee (VANOC) was the first such committee in Olympic history to have been influenced substantially by developments in social media and to have integrated such platforms into its program. These circumstances are explained, in part, by the history of the city and its organizing committee. VANOC was created in 2003, at a time when many of today's most successful social-media companies were just developing. VANOC and the city's population benefited from this shared timeline. The unique cultural contributions of the city's inhabitants are also important to bear in mind. In the middle of its 2010 preparations, the Creative Class theorist Richard Florida described Vancouver as a "first-class creative city" (Florida, cited in Ravensbergen 2008). Among its most successful Web 2.0 offspring are the influential photo-sharing platform Flickr and the social-media management tool Hootsuite. Both of those globally successful social-media platforms were launched from Vancouver as the city was planning its 2010 Games. Thus, Vancouver was a city that was always likely to have digital content at the heart its 2010 Olympic program. The 2010 Vancouver Games were also the first to benefit from the presence of bloggers at the organizing committee level since the early stages of the hosting process. However, the particular politics of Vancouver meant that the new-media community fragmented quickly. As many as six media centers were active during the Games. This collection of communities shows how the city's digital elite was actively engaged around the Games, but in different ways and with different goals.

First, there were what I will call the *Tier 1* media environments—the official media centers, created by VANOC as part of their Olympic Games venues commitment, which were outlined in chapter 6. Next, *Tier 2* media spaces were created and run by the regional provinces; they consisted of two Non-Accredited Media Centres, the British Columbia International Media Centre (BCIMC) and the Whistler Media House (WMH). These two spaces replicated the provision at previous Winter Games, occupying the central square of the city and having a dedicated space in the mountain village. Each of them provided facilities for media that had little or no access to the sporting events, such as the Canadian Broadcasting Corporation (CBC), which had a base in WMH throughout the Games. Yet they also provided an additional space for accredited media to use, should they require it.

The BCIMC occupied Vancouver's main square—Robson Square—and brought together media from diverse organizations, housing studios and press conference facilities. Often the program of events delivered at the BCIMC would overlap with the official media opportunities generated by the Games. For example, one event held at BCIMC was a press conference with California's Governor Arnold Schwarzenegger after he had finished running a leg of the torch relay. Such an example reveals how there are intersecting interests that operate around official Olympic activities and the city-based interest groups who will have protocol privileges that surround the visits of VIPs. Indeed, the value of having secondary, non-accredited media spaces such as the BCIMC and the WMH is precisely in how such venues allow a host city to optimize the value of the Olympic period, particularly in terms of international place-marketing.

Tier 3 media environments also emerged in Vancouver during the Games and these were politically more diverse than the Tier 1 and 2 centers. They were also less populated by professional journalists. Indeed, what distinguishes these three different tiers is the way in which they are organized, which has a direct impact on the types of activity they program and the kinds of people they reach. Tier 3 Olympic media spaces operate outside of any official city or Olympics-based governance structures. In Vancouver, I was directly involved with the creation of two such centers, the W2 Media Centre (set up by W2 Media Inc.) and True North Media House (set up by an independent collective of new-media professionals). The specific facets of these two entities are useful to unpack in a little more detail, and their members occupy the core of what I describe as the *alternative media*.

The World's First Olympic Social-Media Center

The W2 Centre, located in the Woodward's Building in the Downtown Eastside neighborhood of Vancouver, is a cultural organization with considerable historical connections to its community. During the 2010 Games, it created a media center for journalists who were unable to get access to either Tier 1 or 2 media environments or for which, the character of such environments was not appropriate. W2's location brought a unique quality of experience to its program. Located in an area with considerable poverty and deprivation, it was only a few meters from the Olympic Tent Village, a temporary space occupied by homeless people during the Games as an act of solidarity to express their right to be present during Games time and to articulate their dissatisfaction with how the homeless community were being treated in preparation for the Games.

W2's program resonated with the disenfranchised groups. Yet it also endeavored to negotiate a neutral zone during the Games, programming digital art work that was funded by VANOC while also staging events that criticized the Games. This may have been the first time in Olympic history that an organization fulfilled these dual functions during a Games, though this neutral ground may have been seen as a compromise for both of its stakeholder interests. The W2 program was also achieved through a unique cultural collaboration between the Winter and Summer Games, which are traditionally quite separate in their trajectories. W2 worked with one of the regional cultural programs of the 2012 London Games to co-program content that brought the two together. This possibility was afforded by the London 2012 model of producing Cultural Olympiad content, which involved giving a substantial, regionally devolved program to pursue its own vision, simply bearing in mind the Olympic and Paralympic Values.

The W2 story suggests how the Olympic Games program can be brought into closer contact with new communities, which often operate at the fringe of the Olympic industry. It also shows how art and journalism provide a complimentary synergy for fringe activity during the Games. At the opening of the W2 Media Center, the mayor of Vancouver, Gregor Robertson was in attendance and proclaimed it the first Olympic social-media center in history. W2 set up a formal accrediting process that required payment, but there was no expectation that accredited persons would reach a particular outlet. Professional journalists who had not gained entry to the other media centers valued W2's provocative Games-time debates on housing and doping, as these were conversations that were not happening anywhere else in the city during Games time.

The DIY Olympic Media Centre

While W2 provided an alternative media center for journalists to use, the True North Media House (TNMH) collapsed even further the idea that being a journalist was possible for only those who were professionally qualified. The idea for TNMH derived from collaborative work by Kris Krug, Robert Scales, and Dave Olson (all pioneers of new media in British Columbia and early adopters of the Internet) and me. Along with Boris Mann, Krug, Scales, and Olson went to the 2006 Torino Games—held soon after Vancouver won the 2010 hosting rights—and worked with British Columbia to run a series of events, including a meeting of bloggers. These early experiences were the context for TNMH and their experiences of the Torino Piemonte Media Centre brought some insight into what an alternative media center might do—or not do—for citizen journalists whose work was being produced while on the move, mostly using mobile phones.

For a period of time, this community attempted to influence Vancouver's Tier 2 media provision, but the window of opportunity for bloggers to gain accreditation came and went quite quickly; some got through, but many did not. In response, TNMH was created. It became the first genuinely democratized Olympic media accreditation system. To become an affiliated reporter of TNMH, all one had to do was go to the website, download a blank template for a media pass, input one's details and photograph, print a pass, laminate it, and hang it around one's neck. This do-it-yourself media accreditation was, then, a way for citizens to present themselves as reporters. In the strange Olympic world, where a pass around one's neck is the principal mechanism for asserting authority during the Games, this laminated accreditation empowered people to take ownership of the Games and assert their right to access.

This provocative intervention democratized the otherwise exclusive access the media enjoy. It testifies to the imminent challenges posed to the Olympic Games by citizen journalism. After all, to the extent that the

Olympic Games rely on the revenue generated by the sale of media rights to broadcasters, and to the extent that this requires exclusive contracts for the broadcast of Olympic sport and broader Olympic program assets, opening up media access to anyone may jeopardize the Olympic Movement's entire economic base, especially where this involved creating video content. This is the threat posed by citizen journalists, who seek to undermine the processes by which value is attributed to one media outlet rather than another. Yet one may also look at TNMH's proposition as an opportunity, since there is more value in having audiences engage with a brand through the production of creative artifacts than only through the consumption of content. It is in the interest of the Olympic movement to be part of the catalyst for content generation by the public and accept that this will also include content criticizing the Games and their organizers.

Of course, nothing about the Vancouver Games was derailed by the initiative of these Tier 3 media communities, and no financial base was undermined. W2 and TNMH were tiny interventions in comparison with the scale of the official Olympic media structures. However, these acts of asserting media citizenship exemplify the changes that are afoot in the media industries today. When citizens and professionals generate journalism together, the citizens should also be given access to experiences that otherwise are afforded only to professional journalists.

London 2012: An Olympic Citizen Newswire

What took place in the alternative media centers during the 2010 Vancouver 2010 Games informed what happened at the 2012 London Games. In the two-year buildup to the London Games, a similar community of citizen reporters became established around the UK, some of whom were directly involved with the Vancouver programs. The aforementioned collaboration between W2 and the London 2012 Creative Programmers in England's Northwest and Southwest and in Scotland, which had provided investment for the Vancouver 2010 Games-time program, extended into the London 2012 plans. Moreover, a number of cultural organizations got behind the initiative and supported the development of an alternative news network, which became known by the Twitter hashtag #media2012. This collaboration led to a series of conversations and additional programs of activity delivered by the network. For example, a meeting of the #media2012 steering group held in Bristol at the beginning of 2012 was the start of a collaboration toward establishing a Games-time media center (a Tier 3–style venue) near the Olympic sailing venue. Out of the #media2012 community, the #CitizenRelay project was developed, which gave people in Scotland the chance to follow and report on the journey of the Olympic torch across Scotland.

Thus, during preparations for the London Games there already existed a community of international bloggers who had connected at previous Olympics and who were planning to create opportunities for citizen journalism at future Games. For example, as part of the W2 media center program in Vancouver, Alex Zolotarev from Russia took part in a day-long conference titled Fresh Media Olympics at which he spoke of SochiReporter, an initiative (funded by the Knight News Foundation) to build a citizen journalism community around the 2014 Sochi Games. Though this was not indicative of a sustainable, organized network, it did reflect the growing worldwide community of Olympic citizen journalists. Though there was some continuity between the Vancouver, London, and Sochi Games, it was the coincident rise of citizen journalism more widely that really sealed London's fate as the first Olympic Games during which citizen journalism was influential and substantial. Furthermore, important institutions were beginning to support similar initiatives to empower people as journalists.

Alternative Olympic News as Opposition

Outside of the networks described above, citizen journalists were also utilizing social media to amplify antagonistic messages about the Games. As table 10.1 shows, such opposition was apparent was on Twitter. There also were smaller initiatives, such as one opposing the use of animals in the Olympic opening ceremony. (For a more extensive overview of these various campaigns, see Miah 2014.)

Focused activity by citizen journalists is still considerably less visible at the Olympic Games than content and narratives generated by the official, accredited media, but the latter do often amplify the activity of the former. Indeed, many of the campaigns mentioned above were pre-Games initiatives; few of them involved a Games-time delivery plan, as the small number of Games-time tweets reveals. This may indicate a certain kind of resignation over trying to influence Games-time narratives, or it may reveal the duality of the Olympic opposition, which is that it is often directed

Table 10.1

Organization	Twitter Followers, as of 13 Aug, 2012 (day after the London 2012 Olympic Games Closing Ceremony)	No. Tweets during the Olympic Games	First Tweet
@OurOlympics	3,701	829	Our Olympics: a case for reclaiming the London 2012 games http://www.opendemocracy.net/ourkingdom/kerry-anne-mendoza/our-olympics-case-for-reclaiming-london-2012-games @occupylsx @occupylondon @occupywallst #occupy #ows #nhs
@GamesMonitor	972	100	debunking Olympic myths
@CounterOlympics	714	459	Reclaim housing and stop the Olympic Village sell-off on Feb 12 Olympic Village Tent City Vancouver http://vanact.wordpress.com
@BigLotteryRfnd	224	59	Already the DCMS has issued a response to our campaign. They pledge the money will be repaid, but we want details. http://ow.ly/4vU4f
@PlayFair2012	183	3	Get involved in the Playfair2012 campaign—tell Adidas, Nike and Pentland to respect workers' rights http://tiny.cc/vsab4
@DropDowNow	117	1	Drop DOW Chemical as partners for the London 2012 Olympic Games #Bhopal http://www.change.org/petitions/drop-dow-chemical-as-partners-for-the-london-2012-olympic-games-bhopal?share_id=JEYrOwiLbk&utm_source=share_petition&utm_medium=twitter via @change
@BP2012Greenwash	308	43	New twitter to call BP to account for Olympic greenwash. Couldn't be easier really, we just need to RT everything @BPLondon2012 tweets!
@ReclaimOurBard,	238	39	private
@OlympicMissiles	194	0	dummy missiles left "unguarded" during Olympic military exercise http://t.co/OcPnyLyA#stoptheolympicmissiles #olympicmissiles\

toward the politics of the institutions involved rather than toward the idea of sports per se. When the athletes come to the Games, there is perhaps less public empathy for any activism that could jeopardize the athletes' opportunity to realize their lifelong dreams, and perhaps also a degree of respect. This is hard to verify, and it is not true to say that Games are free of protest—far from it. For instance, during the Vancouver Games, there was marching in the streets of the city on the day of the opening ceremony. In 2012, on the day after the opening ceremony in London, more than thirty organizations gathered at Mile End Park in that city, organized by the Counter Olympics (@CounterOlympics) Network, to march in protest of the Games.[2] Similar marching took place at the Rio 2016 Games, though mostly in protest against the Brazilian government and against diverting resources away from critical social needs toward the Games. Although these events were noticed by the media and reported, they did little to derail the core stories that all rights holders were politically committed to telling. If there is a single message to be drawn from the story of citizen journalism at the London Games, it is that the politics didn't always oppose the Games but always asserted media citizenship. London 2012 signaled Olympic fans' desire to contribute to producing the program, not just to consume it. Of course, the softer form of citizen journalism as social-media content is a quite different matter. In this case, it is difficult to deny that the volume of activity generated by citizens is fast becoming the most influential news content to circulate about the Games, but how it becomes organized to have an impact on the prominence of the mass media is still unresolved.

Sochi 2014: From Ambush Marketing to Ambush Media

The disruptive potential of alternative media represents a common feature of media change more widely. The creation of any new newspaper, magazine, television channel, or radio station—if successful—is a disruptive influence within media culture. Thus, the disruption brought by citizen journalism through social media may be more in degree than in kind. Yet the harnessing of social media by citizen journalists to intervene in the centralized storytelling of an Olympic Games is a pivotal moment in Olympic media history. Moreover, there are instances in which such interventions have been significant—notably in the case of the 2014 Sochi Games, during which there was a debate about the rights of LGBTQ Russian citizens.

As February 2014 approached, campaigners wrote a letter to the CEOs of all the major Olympic sponsors asking them to take the following actions:

- Individually or collectively, condemn Russia's anti-LGBT law.
- Use your Olympics-related marketing and advertising—both domestically and internationally—to promote equality during the weeks leading up to and during the Games themselves.
- Ask the IOC to create a body to monitor serious Olympics-related human rights abuses in host countries as they occur; and
- Task the IOC with ensuring future Olympic host countries honor their commitments to upholding the Olympic Charter, including Principle 6 which forbids discrimination of any kind.

When the sponsors gave no response, the activists hijacked a McDonalds Twitter campaign hashtag—#CheersToSochi. That resulted in prominent syndication of the alternative content. Campaigners set up a website that mimicked the layout of the McDonald's campaign, using similar layouts, colors, and titles, but instead contained a slightly modified text detailing the abuses and identifying sponsors as complicit in supporting them by not withdrawing from their relationship with the Sochi Games.

In traditional ambush marketing, a brand seeks to position its identity into the space of another brand. The aforementioned forms of disruption are more accurately described as ambush media campaigns. Their merit is located principally in being able to insert a subtle alternative message by hacking into the design elements of the originator's campaign. Unlike ambush marketing, which typically implies one brand or product seeking to gain visibility, these forms of ambush media are not about a brand or making money, but are about using social media to distort media culture. Examples of ambush media campaigns also took place around the 2012 London Games in relation to the company British Petroleum (BP), which received criticism ahead of the Games. In this case, the Campaign for a Sustainable Olympics (CAMSOL) created a website which emulated the design features of the London 2012 website, apparently conveying a sincere message about BP's being removed from the Olympic program. Yet on closer reading, it was apparent how this was completely fake and that the proposal to drop BP from the domestic sponsors portfolio was not, in fact, going to happen.

Examples of ambush media are important since a crucial element of the debate about citizen journalism's value in the context of the Olympics concerns its impact on the wider Olympic media culture. It is not yet apparent whether these different journalism populations are all complementary. To some extent, the objectives of each media population are determined by their access to the Games program. Thus, only Tier 1 media have access to the sports, and they tend to spend the entire Olympic period delivering assignments related to this aspect of the Games, perhaps never even leaving the Olympic sports news bubble. The Olympic machinery serves them by ensuring that the athletes are positioned in front of the cameras in the order of the importance of those to whom the images will be available: first rights-holding television stations, then non-rights-holding news television, then accredited journalists. The work of accredited photographers or writers is slightly different, but even these areas of content generation operate within similar restrictions. For text-based reporting, the privileged access takes the form of the "press Tribune," a special stand where writers can sit to watch the action and write about it during the event, the importance of live witnessing being still a crucial element of the journalist's professionalism. These arrangements leave little scope for Tier 2 or Tier 3 reporters to get near to the sport, so they must focus on something else, and that is exactly what they do.

In any case, the journalists accredited by Tier 2 and Tier 3 media centers are not really focused on the sports at all. Although some of them are sports journalists who were not able to secure access to Tier 1 centers, many of them have interests that extend beyond sports to the host city, politics, or tourism. Typically, Tier 2 journalists cover everything related to the Olympic program except sport, and their daily deadlines are supplied with stories from the city and its stakeholders. Tier 3 reporters covered more controversial content, often what I describe here as the focus of alternative media, but even this is a mixed community. Many citizen journalism projects around the 2012 London Games simply provided members of the public an opportunity to tell Games-related stories themselves. For example, the Whose Olympics? project aimed to "produce a collaborative documentary telling the story of how London's parks and open spaces are changing in 2012 as a result of the Olympics, and how these transformations are affecting people's experiences of their local areas and of the Games."

Together, the three tiers of the Olympic media infrastructure are indicative of wider changes in the relationship between media institutions and society. In today's world of user-generated content, media citizenship has grown. Along with this, there is a greater demand for access to spaces that were historically just reserved for the professional media. The consequences of this are difficult to foresee, but in the case of the Olympic Games there are a number of outcomes to these trends. For instance, these changes create a greater desire from the accredited media to extend their rights to encompass the non-sporting elements of the Olympic program. Indeed, this may resonate with the IOC's Agenda 2020 plans to create a "360° Olympic program" experience for fans that will go beyond what happens in the venues during the Games. Alternatively, the Olympic industry might find a way to promote citizen reporting in a way that aligns with its wider goals. Indeed, during both the London Games and the Sochi Games the IOC and the organizing committee held dedicated press conferences just to talk about the volume of social-media activity. Thus, on one level there is certainly a sense in which the "long tail" of the media population assists the Olympic movement in its goals.

The long-term consequences of these changes remain unclear, but the expansion of user-generated content has already given rise to a new era of media operations at the Olympic Game. Media outlets now staff the Games with new roles. Positions such as Web architect, bloggers, and content curator, and even the specialized tasks of developing data-driven journalism initiatives, are novel roles for Olympic reporters, and for reporters more generally. Alternatively, one consequence of an open media environment might be even tighter restrictions on the public. For instance, in Vancouver the privilege gained by TNMH accreditations was due in part to the inability of Games-time systems to adequately distinguish between types of media accreditation (or their indifference to them), coupled with the desire of non-sports venues to draw in other media. Indeed, it isn't clear whether TNMH accreditations were getting participants into venues where they should not have been allowed, though it is apparent that what institutions recognize as important media is controlled solely by a device we call a press pass, even if that pass says nothing about the community that is reached by that individual or the team holding it. Yet the interesting feature of the TNMH phenomenon was the idea that all one had to do to have the access of a professional journalist was to assert oneself as such. Alternatively, in the same way that separate contracts have been set up with YouTube, an expanding media population might lead the IOC to set up alternative media centers for the greatly expanding network of reporters who want to cover non-sport elements of the program.

The value of content generation for citizen journalists at the Olympic Games is measured not in column inches but by its capacity to have engaged people who might otherwise be mere spectators or passive tourists. Community media (more specific, participatory media) emerges from a lack in other forms of media production, which are perceived to be failing in their social responsibilities to communicate information, or because they are governed by a political agenda that undermines the integrity of the content. However, the more persuasive reason for why community media is necessary is that the production of media artifacts *creates* communities. The desire to share opinions and knowledge is a powerful motivation that explains why community media exists. It is an integral part of how one defines citizenship, freedom of speech and what it is to be human.

While the expansion of media technology has narrowed the digital divide, there remains a growing *digital literacy* divide that community media organizations can help to address. At the same time, we now operate within an *attention economy* where the biggest challenge for media outlets is the small window of opportunity through which they have the chance to capture people's interests. It is said that the average life of any social-media artifact is three hours, after which it is highly unlikely to garner much interest at all. In part, this has changed the role of community media organizations where an increasingly important part of their job is to curate the media output of community members, rather than provide a media production service for a community.

The alternative media communities at the Olympic Games are composed of people who challenge the privileged position of traditional media organizations. The term *ambush media* implies piggybacking on the intellectual property generated by the financial power of traditional media and turning the camera lenses back on these institutions to reveal which stories remain untold.

There are many ways in which this latter phenomenon can be observed at the Olympic Games. For example, it includes the simple act of allowing accredited journalists to enter a non-IOC regulated media space where which they will learn about less visible Olympic activities. It also includes the fact that non-professional journalists can broadcast and write about the Games—as is true of the Tier 2 and 3 media centers—thus providing opportunities for a wider range of questions to be asked during the actual event. The functions served by Tier 3 reporters at the Games is akin to what Beckett and Ball describes as "outsider journalism," which they consider is crucial to "challenge media and power" (2012, p. 156). They say that "there is still significant public support for that kind of "irresponsible" journalism that catches both the authorities and mainstream media off-guard" (ibid., p. 146), and what has taken place around citizen journalism at the Olympic Games is a lot like this. On one interpretation, it is often a small group of creatively enabled individuals who have the skills to make an impact online. Indeed, at the 2010 Vancouver Games, the social-media story of W2 and that critical creative community was made into a feature-length film titled *With Glowing Hearts*, which provides the alternative version of those Games. Though it isn't likely that the film will end up in the IOC's archives for Vancouver, there is merit in concluding that, were this to happen, the Olympic movement would be in a far better position than it is now to assert itself as a social movement, rather than just a mega-event.

In sum, the rise of citizen journalism at the Games allows the Olympic movement to more effectively attend to its underlying values and its priorities as a social movement. Social media provide a vehicle through which Olympic organizers can empower citizens to be more active in shaping the narrative of each Games and to ensure more meaningful ways to take part in the Olympic program. For this reason, it is crucial that the Olympic industry work more closely with such communities at a hyper-local level to maximize the opportunity that is brought to the Games, drawing upon the prevalence of social media to do so.

Conclusion

This book has explored the digital ecosystem that surrounds sports in the 21st century, which show how they have become deeply intertwined with complex digital configurations. These systems are so vast now that it is reasonable to ask whether there is any future for physical spaces within sports, but the growth of e-sports shows how Sport 2.0 leads to a re-imagination of space, place, and physicality. Indeed, my central proposition throughout this book is not that the future of sport will be free from physicality, or that digital corporeality will replace physical activity. Rather, I have argued that it is necessary to consider new designs for their inter-relationship, as spectators, athletes, and officials become components of a wider biodigital infrastructure, much more connected than ever before. In the same way that people may think of themselves as intimately connected to their mobile devices, be those watches, bracelets, or phones, the future of sport for athletes, spectators, and officials will be characterized by the integration of digital technologies. It is now time for sports to abandon the distinctions between physical and digital ways of being. In sports culture the evidence of this is overwhelming. Indeed, even the idea of wearable technologies may soon become outmoded as ingestible technologies emerge. For example, in September 2015 the US Food and Drug Administration awarded Otsuka Pharmaceuticals and Proteus Digital Health the first license for a form of medication with a built-in sensor. The technology allows a mobile device to connect with the medicinal product as it travels through the patient's body:

When ABILIFY with the embedded ingestible sensor is taken, the ingestible sensor sends a signal to the wearable Proteus patch after it reaches the stomach. The patch records and time-stamps the information from the ingestible sensor in addition to collecting other patient metrics, including rest, body angle and activity patterns. This information is recorded and relayed to patients on a mobile phone or other

Bluetooth-enabled device, and only with their consent, to their physician and/or their caregivers. Patients view the information using a secure and local software application on their mobile phone or device. Physicians and caregivers view the data using secure web portals. (Proteus 2015)

Such digital systems are not just technological; they are also cultural, social, political, aesthetic, and moral. Their ubiquity allows organizations, individuals, and communities to re-negotiate how they exist in the world, to the point where one may now speak of a digital ideology that pervades all aspects of life. The consequences of this new world view are manifold and sometimes contradictory; however, as more data is generated, yielding insights into our lives, there are growing concerns about how the data will be used, who owns it, and what we might be subjected to as a result of its being managed by commercial companies.

We are only beginning to see the consequences of the shift toward digital societies, and many of the values of today's sporting culture will have only limited relevance for the new digital world. High performance, the increased precision and predictability of events, and the increased parity among competitors are causing sports to radically revisit their values, returning to a more ritual-based culture, rather than one that is focused on records and results. After all, in a world where first place and eighth place are so close that the average spectator can no longer tell who has won, there may be other reasons to watch sports than to see who is the victor. In such a world, digital mediation will become even more crucial and will become a way of disturbing and tweaking the field of play. Consider what takes place in the new sport of Formula E auto racing, in which the performance of a car can be affected by the audience's enthusiasm. According to the Formula E website, fans can give their favorite driver a performance boost by voting before the race; "the three drivers with the most votes will each receive one five-second 'power boost,' temporarily increasing their car's power from 150 kW (202.5 bhp) to 180 kW (243 bhp)." These novel propositions show how the assumptions we make about the relationship between athlete and spectator can be changed.

The opening chapters of the book located the discussion about Sport 2.0 in a wider philosophical context that draws on play and game theory. By making sense of the connections between digital culture and sports culture—along with their manifestation as Olympic culture—one can see how this transformation into digitally mediated sports is occurring. Sports need digital technology in order to advance as cultural pursuits, especially as audiences and physically active people grow to expect a data-driven life-style. One key area of common ground within this relationship is in the rise of digital gaming, which reveals—like sports—a desire to forge alternative worlds, and is further evidence of why sport and digital lives are moving in similar directions.

The specifics of these shifts are also apparent in how digital technology supports elite athletes in the pursuit of their achievements. Whether for performance feedback or for practice through simulation, digital solutions offer effective ways for athletes to enrich their knowledge about which areas of their performance may need improvement. Officials can draw on technology to make better judgments, and athletes can develop their public personas by utilizing social media. The rise of serious gamers as elite ath-letes is a further reason for why traditional sports—and their athletes—need to come to terms with the digital turn in their industry. After all, sporting celebrities are very well placed to exploit their audiences' interests, but they risk being marginalized by the burgeoning e-sports industry, which is devel-oping its own markets, its own systems of monetization and broadcast, and its own distinct brands. If the game-development community remains absent from traditional sports, then sports face a future in which they are further marginalized and less relevant to younger populations, for whom the location of those new experiences—online—is more familiar than tra-ditional collective spaces, such as gymnasiums and sports halls. The same is true of spectators, who have ever-growing demands for personalized, unique point of view experiences. Spectators like to take photographs and videos at live events because doing so is a creative act that leads to further investment in the experience—it personalizes the memory and allows a fan to feel more a part of what took place. This "eyewitness" aspect of sports spectatorship is why the future of digital spectating is not bleak. Rather, it is likely to be characterized by hyper-mediated encounters with extraordinary levels of technological mediation.

Three notable shifts apparent in the Olympic Games explain why the e-sports community is so tied to online and mobile media. The first is the shift within the professional media, which is changing in demographic and in terms of what they do at an Olympic Games. The diversification of platforms on which media outlets locate their content reveals a new hierarchy of media power, in which third-party social-media platforms such as Twitter, Google+, Facebook, and Instagram have become crucial to engaging audiences. In turn, journalists are required to figure out how to use these platforms effectively; writers must be photographers; and everyone must be a video journalist. It is also apparent that the larger an event becomes, the more likely it is to draw journalists from other areas of interest, not just sports. The rise of the non-accredited journalist at the Olympic Games and the growth of citizen journalism are further indications of these changes. Because the Olympic Games rely on the exclusive access to content afforded to the accredited media, the challenge for the IOC is to enable this new population of reporters to produce content that does not compromise these arrangements but can ensure that the Games deliver the most comprehensive coverage that allow them to assert themselves as a social movement, not just a sports event. Embracing alternative media is crucial to this, but the Games have yet to find a way to embrace this kind of journalism without over-managing it.

In the first decade of the revolution brought about by social media, the world has seen Twitter contribute to the so-called Arab Spring of 2009, the emergence of WikiLeaks, the emergence of augmented-reality devices and wearable cameras, and an explosion of e-sports gaming. Where does this leave society? How can these changes be for the better, rather than just more of the same? In April 2016, Twitter signed a contract with the NFL to live-stream football matches within their micro-blogging platform, signaling a huge shift in what kinds of experiences people can expect to enjoy within the Twitter platform. The agreement is also a powerful statement on the future of television, which further reinforces the potential of the e-sport model, and the expansion of mediated sport into novel environments. In 20 years from now, television as we know it today may no longer exist.

In the opening ceremony of the 2012 London Games, one of the most poignant moments was the appearance of Tim Berners-Lee at the end of a portion of the ceremony that focused on the roles of social media and digital technology in the lives of young people. That portion of the ceremony began as homage to British popular culture, taking viewers from the early 1950s right up to the present. Berners-Lee's appearance at the head of the stage, sitting behind a computer, became iconic of the ceremony and of the 2012 Games, widely discussed as the first social-media Olympics. The artistic director of the ceremony, Danny Boyle, was widely thought to have promoted a series of political ideas through the ceremony, one of which was epitomized by this brief segment.

The ceremony in London concluded with a graphical display of a quotation from Berners-Lee, "This is for everyone"—words that were also sent by Berners-Lee while in the stadium via Twitter. The quote conveyed the sentiment that the Internet should be free, open, and, above all, beneficial to the whole of humanity. At a time when the British government and others elsewhere were investing in greater control of the Internet, this was a particularly important statement to make to the world. Berners-Lee's full tweet confirms the underlying values of this section: "This is for everyone #london2012 #oneweb #openingceremony @webfoundation @w3c." His highlighting of the World Wide Web Foundation and the World Wide Web Consortium as organizations that champion a free, open, and accessible Web reinforced the message.

This message provides a sound basis on which to conclude this book. The world of elite sports pivots on two values that are often in opposition: competitiveness and cooperation. In part II of the book, I dealt principally with the competitive side of sports, the utilization of digital technology to achieve competitive or regulative improvements. In part III, I focused on the opportunities afforded by digital technology to promote cooperation and the sharing of sport experiences. For decades, sports have thrived as a result of successfully managing these two principles to promote excitement and anticipation within sports, while also generating increasingly large audiences and more varied modes of participation.

The growing population of digitally engaged citizens may simply be evidence of the involuntary shift from an analog to a digital society. It may not be an indication of a more computer-literate society, or even of a society that has found digital technology to offer exciting new modes of personal experience. Blogging, photo sharing, video sharing, and virtual reality are indications of such change, but they risk also being indicative of more widely networked trivia rather than of content that can change the world.

Expectations of digital culture have always been framed in such a way, but is it reasonable to have such high goals for new technology? Indeed, if social change is expected of digital technology, then what kind of change is sought? The most celebrated uses of digitally enabled empowerment have focused on forms of political activism—literally attempts to transform political systems. Yet change can occur every day and have important political consequences, though they may not involve disruption to power relationships. Alternatively, the use of online chat can dramatically improve a person's relationships, particularly in an increasingly mobile world.

It is also apparent that being digitally engaged cannot just involve being part of a developer community, or being a citizen journalist. To this extent, it is both naive and undesirable to isolate digital participatory culture from society and culture generally, as if only participation in the former is an effective mode of social change. This acknowledgment may require that we set an optimal ratio of digital citizenship, say shifting from a situation in which there are 2 percent producers and 98 percent consumers to one in which there are 20 percent producers and 80 percent consumers. Alternatively, one might propose an ideal digital era in terms of a percentage of time online. For example, perhaps an ideal scenario is one in which 100 percent of users dedicate 10 percent of their time online to production and the remaining 90 percent to consumption. This may seem simplistic, but the point is that if 100 percent of digital citizens were to spend 100 percent of their time producing content there would be no chance to consume or interpret what is available, so we must be able to assert some claims about an optimal ratio.

It may also be necessary to reassert how the movement of information corresponds to social action in a digitally mediated era. For example, could we envisage a time when it may be concluded that an optimal number of blogs or media outlets has been reached, beyond which further information would lead to oversaturation or even disengagement with the agencies of such information? Might we make such a claim about today's media? An initial response to this would be to draw attention to the different roles of information. For instance, information may be important for the purpose of present-day debate, relevant to democratic processes. Yet it might also have historical value, the importance of which will not be fully realized until some time has passed. Arguably, society has become so focused on the former that the value of the latter has been overlooked; but why should information be of use only in the present? Digital platforms such as Snap-Chat, which are valued because the content generated disappears soon after it is shared, may be a resistance to the idea that all digital artifacts will be around forever, but it may also jeopardize our ability to understand history.

Given the rise of digital technology, the growth in virtual-reality systems, the greater public engagement with and use of digital systems, and the expansion of all these elements within the world of sport, I return to the original conceptual foundations of this book and ask whether there is now any need for games to exist in the physical world. In view of their burden on the world's resources, why not migrate sports entirely into the virtual world and wave goodbye to the physical-world sports that currently dictate the conditions of elite and amateur participation? After all, trends toward virtualization are apparent in all other walks of life. Never again would rain stop play in golf or cricket, or be a burden on the world's water resources. Never again would a city be burdened by a need to create "white elephant" stadiums to house fleeting sports events. Instead, the world could pool its resources and invest in developing radical digital solutions, the benefits of which would trickle down into society at large—something like a space program for the sports world. Athletic teams would not need vast amounts of funds to travel across the world to compete in events, promoting a more democratic participation. The unpredictability of the natural world need never again compromise the integrity of a competition result, as powerful simulations could precisely stabilize all irrelevant environmental variations, which lead to unfair inequalities.

Even while writing this, recognizing that these possibilities articulate many of the values sports endeavor to promote, I can imagine the responses of sports administrators and enthusiasts around the world. For instance, the Olympic Games trade on the social value of bringing the world together in one place at one time. It is for this reason that the IOC resists even the idea of two neighboring cities' sharing the burden of the Games. But what would really be lost if sports were to take the logic of virtual worlds to an extreme? Sports would still require considerable international collaboration—perhaps even more than they now do—to produce competitions entirely within virtual worlds. There would be no sense in which sport would lose its role as a social movement, aimed at fostering more effective geopolitical relationships. Indeed, if one considers the raging debates about the management of the World Wide Web and the Internet generally, this is perhaps one of the most crucial geopolitical conversations taking place today, requiring considerable diplomatic negotiations.

The loss might be in terms of the social aspects of sport—the gathering together of spectators, the journeys through cities to venues, the experience of being there. Such factors might deserve preservation were it not for the fact that the world I imagine would operate by very similar principles. Sports in digital worlds more than adequately match the immersive encounter of physical-world experiences. Indeed, they might even be more immersive. A publicity line for the new OLED television showcased at the

2013 Consumer Entertainment Show was "better than real life." We are likely to hear that phrase again and again in the next two decades.

In regard to who will win the communication battle in sport, I noted earlier that social media and perhaps even citizen journalism do not jeopardize the foundations of the rights-paying media. Thus, one may conclude, as did Frost (2011, p. 325), that the "big players will eventually dominate on the Net" and that social media or citizen journalism will never occupy the same position as those big players. However, many new media networks can and have become big players in that world, and the big players' social-media platforms are gatekeepers for the professional media. In that sense, the political configuration of communication has changed. Whether or not this empowers more citizens is a very different question, but conversations about the rise of the hyper-local experience are important to bear in mind. Though one may not want to hear what anyone has to say, one does want to hear one's best friends' experiences, and even Facebook's Graph Search engine attempted to engage with this shift, albeit unsuccessfully. Unless the large media companies can generate this kind of personal attention, their position is not assured, at least not in the day-to-day social newsgathering that many people seek to enjoy their leisure time online and off.

A final consideration about the next stages for digital sport relates to the growing bio-digital interface, the consequences of which are likely to transform many of the debates already considered here. The implications of this fusion between biological and digital systems extend from the rise of wearable technologies, such as Google Glass, but the crucial differences here is the degree to which the technology becomes an integral part of human biology. The early indications of such interfaces can be found in advanced prosthetic devices, a symbol of which was the inclusion of Oscar Pistorius in the 2012 London Olympic Games This was the first time that a prosthetically enhanced athlete had competed alongside so-called able-bodied athletes during an Olympic Games, and his appearance was widely discussed as the rise of the bionic athlete, whose successors will challenge the status of the biologically constituted athlete. The Cyberathlon Games held in Zurich in 2016 are a further indication of this technologically enhanced future, and, as with everything else discussed in this book, the realization of such an event is heavily reliant on digital data systems.

There are good reasons to suppose that this trajectory toward bio-digital integration will continue and that athletes will make digital technology a bigger part of their biology. Indeed, prototypes are beginning to reveal how physical interfaces and sensory experiences may change, giving rise to new ways of experiencing the world. For instance, Kevin Warwick has designed chips that permit neurological exchange to take place—a kind of telepathy. The biodesigners James Auger and Jimmy Loizeau have created a telephone tooth implant that would allow people to interact with others without the need for any external interface.

If all of this seems a long way from today's world of sport, consider how technology has already changed athletes' biological capacities by modifying the world around them. For instance, piezoelectric dampening devices that reduce the impact of the piste on joints have been used in skiing for twenty years. Even these external technologies affect an athlete's internal biological equilibrium. In all aspects of research into digital technology and sports, there are countless questions that have yet to be answered. Some of them require urgent monitoring. For example, how does access to digital sporting experiences occur across different demographic groups? Though a lot is made of the rising presence of women in digital gaming, is the sports genre evenly distributed? Taking into account the prominence of soccer games in major tournaments, coupled with the overwhelming presence of men's in fantasy sports experiences (Howie and Campbell 2015), it would be premature to conclude that digital sport will be gender neutral. Alternatively, the rise of data-driven exercise raises serious questions about the management of data generated by such exercise and who will have access to it, not to mention how such activities as match fixing may operate within a world of e-sports.

Alternatively, as more of our personal histories are transformed into data, there is a growing need to advance a right to personal data ownership and, in particular, the need for a universal export of our data from one platform to another. Otherwise, the long-term positive consequences of the Big Data revolution will be swallowed up by closed, commercial systems, which will undermine our capacity to make sense of our health without paying for the privilege. While recognizing a right to own our health data is far from becoming a social reality, there is a point at which the public good associated with data should outstrip its commercial value. The demands for greater ownership and agency in our data-driven economy also resonate with the values of citizen journalism. What is at stake with the shift toward an expanded media operation at the Olympic Games is the transformations occurring in the practice of journalism, driven by the new capacities of citizens to generate their own content and curate their own stories. Equally, the move toward "backpack journalism" (Edgar 2013, p. 1208), or what Boyle (2006, p.138) earlier described as the "wireless sports journalist," articulates the changing needs of journalists at an Olympic Games. The role of a media center is becoming less relevant as more journalists are self-contained production facilities. All they need to do their work is a electrical power supply and an Internet connection. As time goes on, accredited media at an Olympic Games are gradually blending into the crowd of citizen journalists, provoking debate about what should distinguish professional media in this new era of networked journalism. Each of these examples reveals how the sports world would be richer if it were able to adopt more citizen-led goals in its pursuit of economic sustainability. The best way to engage an audience or a readership is to make them part of the story's creation.

The grassroots-led e-sports gaming movement is a reference point here, as an alternative model, but it also risks selling out to big sponsors along the way.[1] Moreover, there is evidence that traditional sports are beginning to appropriate e-sports; for example, in May 2016 the English football club West Ham United signed its first e-sport player. Nevertheless, consumer-driven live streaming of broadcasts may present a model that is—for the Olympic Games and other events—a more effective way of ensuring that the media contracts around such events will, as Kidd (2013) would want, make a wider contribution to advancing sport as a social movement, rather than just a series of competitions. Central to these issues is the allocation of intellectual property rights, which become increasingly compromised as data and content shift freely from one place to another in digital spaces. In the world of e-sports, this is particularly explosive, as the main sports federations have yet to consider their relationship to titles which may be framed around sports that they seek to promote and administrate (Burk 2013). In the future, one might envisage sports federations replacing disciplines from the Olympic Games with digital equivalents, but this will require organizing themselves to ensure that the e-sports gaming community don't just turn their backs on organized sports and set up their own mega-events. After all, it doesn't appear that e-sports will be reliant on achieving sports recognition any time soon, even if doing so would bring considerable social benefits.

Not all forms of digital innovation need erode the importance of the physical. In fact, some examples can enrich the physical world, as might be said of Nike's Chalkbot, a robot vehicle that spray-painted yellow "graffiti" onto the route that was taken by the Tour de France riders. The graffiti consisted of messages sent by audience members using their mobile phones. Creating physical manifestations of digital messages and integrating them in the field of play was remarkable on numerous levels from the perspective of sports production, but the key message here is that creative, high-technology solutions can be delivered through simple, familiar technological means—in this case, SMS messaging. As more and more technologists become involved in the production of Sport 2.0, this is an important aspect of our future that is worth remembering, as it will ensure that digital systems improve social justice, rather than exacerbate inequalities.

Notes

Introduction

1. The 2010 Vancouver Games also had a large amount social-media activity and were claimed to be the first social-media Olympics. However, the 2012 London Games featured more integration of social media and took a more holistic approach. The organizers of the London Games also foregrounded social media in the opening ceremony.

Chapter 1

1. Critics of the Olympic Truce draw attention to its failures—during each Olympic Games, conflicts continue. However, this would be a naive rejection of its value, the consequences of which are felt more widely in the network of actors who work within these organizations, even if the ultimate goal is not achieved.

Chapter 2

1. Bitcoin is an online payment system, created in 2008, that allows peer-to-peer transactions.

2. Some scholars have also criticized this premise, arguing that local indigenous populations considered her selection as final torch bearer tokenistic, playing to an international audience but failing to address tensions between local communities.

Chapter 3

1. For a rich analysis of the interpretive frames around analyzing close finishes, see Finn 2014.

2. In Baldwin 2012, Peter Huerzeler of the sports timing company Omega explains that competition venues are not built to millimeter precision, which is the degree of

difference that measuring to thousandths of seconds would distinguish. For example, a swimmer in one lane of a pool may actually swim a different distance than another, because of slight differences in the pool design. On this basis, measuring success in millimeters of difference would be unfair.

3. On the Brazuca ball designed for the Brazil 2014 World Cup, see Goff, Asai, and Hong 2014.

4. The Italian cyclist Luca Paolini was reminded by the organizers of the Tour de France that using a phone is not permitted during the race after a Twitter user captured an image of him doing so (McMahon 2014).

Chapter 4

1. Zombies, Run! immerses runners in an action-packed game. It is designed to make the outdoor running experience more engaging, by creating an interactive audio narrative around the experience. Players use headphones to hear the story play out as they run, while the game creates a story around the run's goals and missions. The player has to run to escape zombies, who are chasing them around their running route.

2. To complete the circle of media blame, in 2005, the Red Lake High School Massacre was briefly blamed on the film *Elephant*, as it was viewed by gunman Jeff Weise 17 days before the shooting (Lester 2006).

Chapter 5

1. Movies such as *Zidane* (2006) and *The Swimmer* (2012) make us aware of the other ways in which audiences might experience what is meaningful about sport.

2. In May 2014, Google acquired Quest Visual, the maker of Word Lens, for the purpose of incorporating Word Lens into its Google Translate service.

3. As of December 2014, the broadcast rights revenue for the period 2013–2016 was valued at $4.1 billion (Owen 2014).

4. In actual fact, the Olympic Games maintain a "clean" space within the arena itself, so that only the Olympic rings are the visible symbol for television viewers, but most sports events use the arena to drive revenue.

Chapter 6

1. These include the Main Press Centre and the International Broadcasting Centre (sometimes described together as the Main Media Centre) and the Venue Media Centres within various competition venues.

2. Broadcasters provide approximately 4 percent of all Olympic revenue sources; sponsors provide up to 18 percent.

3. A good example of this is Channel 4's 2007 program "Unreported World: China's Olympic Lie," which explored the lived experiences of foreign reporters operating within the new legislative rules in China.

4. For more on the Space Hijackers, see Gilchrist and Ravenscroft 2013.

5. Earlier examples of the art of cinemagraphs can be found in the work of Duane Hopkins, whose *Sunday* was shown at the Abandon Normal Devices festival.

6. Clearly this example could compromise patient-doctor confidentiality, but I propose this precisely to emphasize the idea that, in an athletic context, that relationship is already jeopardized by the commercial nature of the endeavor.

Chapter 7

1. The accredited figures are from a personal communication with the IOC's Head of Media Operations, Anthony Edgar, whose role includes overseeing the media accreditation lists.

2. Yet, as I will argue later, the increasing close relationship between the NAMC and the rest of the Olympic delivery infrastructure means that some access to official Olympic program has been possible. Examples include access to medals plaza events at the Winter Games and access to Cultural Olympiad activities, which are marginalized from international media coverage.

3. In some cases, even an accredited person with complete access may require an "upgrade card" for specific high-profile or special events. For example, a reporter desiring access to the opening ceremony may need a special sticker on his or her accreditation to gain entry to the media zones.

4. Olympia is about five hours from Athens by road, and the organizers may not have expected many accredited journalists to make the long journey there.

5. At the Athens 2004 Zappeion Media Centre there was also some presence of the Organizing Committee for the Olympic Games, but only from its Cultural Olympiad program.

Chapter 8

1. The guidelines were developed by the IOC's Head of Media Operations, Anthony Edgar.

2. Some critics have argued that the mere existence of guidelines restricts individual freedom of expression, but my argument means to challenge this idea.

Chapter 9

1. Two years later (February 5, 2015), the number of subscribers remained at approximately 51,742, with 15,683,645 views.

2. Despite first publishing content on the channel in 2008, the channel was created on March 29, 2006.

3. It is worth noting that the IOC created its channel on January 10, 2006, just two months before London 2012 opened its channel, suggesting that London would be the first Games at which an integrated YouTube policy across stakeholders would be apparent. In part, this may explain the dominance of London in the IOC's assets, but, of course, the fact that the previous Summer Games had been held in China may also have had a bearing on its YouTube presence during the Beijing Games.

4. These numbers are taken from Alex Balfour's first post-Games interview, which took place on the Guardian's podcast Tech Weekly with Aleks Krotoski in October 2012. They are also available on a Slideshare presentation published September 10, 2012.

5. This statistical claim needs to take into account that the unit of comparison is skewed, since the Olympic Games is an *event* that takes place over 17 days. NBC sport also states that the most watched *show* in US history is the 2015 Super Bowl, reaching 114.4 million viewers on a single evening (NBC Sports Group 2015). In comparison, the closing ceremony of the 2012 London Games attracted 31.4 million viewers for NBCUniversal. The other top eight most-watched shows include the preceding six Super Bowl games and, in seventh place, the 1983 finale of the long-running series *M*A*S*H.*

6. The BBC did rely on its @BBCsport account, which had considerably more followers than either of these dedicated Olympic/Paralympic accounts.

7. The absence of any mention of the 2012 London Olympic and Paralympic Games is interesting. Perhaps the controversy it generated among the art community explains this distancing, or perhaps there was no political will to articulate that relationship after the Games were over. Either way, it seems clear that the Olympic brand has no currency for arts producers once the Games are over.

Chapter 10

1. The collaboration was led by Debbi Lander, who leveraged interest from the newly formed 2012 pre-Games festival called Abandon Normal Devices (which produced the co-curated program) and from DaDaFest (the UK's leading disability and deaf arts program). AND's founding partners, Cornerhouse in Manchester and the Foundation for Art and Creative Technology, supported initiative; FACT also sent its community media unit—TenantSpin—to help set up the Games-time studio.

2. The organizations that signed up for the march were listed by the Counter Olympics Network on its website as Action East End, ALARM, Athletes Against Dow Chemical's Olympic Sponsorship, BADHOC, Black Activists Rising Against Cuts, Blacklist Support Group, Bread and Circuses, Brent Trades Council, Camden and Islington Unite Community Branch, Coalition of Resistance, Communities against the Cuts, Counterfire, Defend the Right to Protest, Disabled People Against Cuts, Drop Dow Now, East London Against Arms Fairs, G4S Campaign, Games Monitor, Grunts for the Arts, Hackney Green Party, Hackney Trades Council, Hackney Woodcraft Folk, Haldane Society of Socialist Lawyers, Haringey Trades Council, Islington Hands Off Our Public Services, Islington Trades Council, Jewish Socialist Group, Lewisham People Before Profit, Lewisham Stop the War, Lewisham Trades Council, London Green Party, London Mining Network, Netpol, Newham Monitoring Project, No Games Chicago, Occupy London, One Law For All, Our Olympics, Partizans People and Planet, Save Leyton Marsh Campaign, Socialist Workers Party, Space Hijackers, Stop the Olympic Missiles, Thurrock Heckler, UK Tar Sands Network, Waltham Forest Trades Council, War on Want, and Youth Fight For Jobs.

Conclusion

1. Prize money is generated by community donations. In 2014 alone, the prize money for the International DOTA 2 competition reached $10,931,103 (e-Sports Earnings 2015).

Bibliography

Ahmed, Murad. 2008. "Facebook Fans Lost for Words in 'Scrabble' Row." *The Times*, January 17 (http://www.thetimes.co.uk/tto/technology/article18).

Akins, A. S. 1994. "Golfers Tee Off into the Future." *Futurist*, March-April: 39–42 (https://www.questia.com/magazine/1G1-15266290/golfers-tee-off-into-the-future).

Allan, Stuart, and Einar Thorsen. 2009. *Citizen Journalism: Global Perspectives*. Peter Lang.

Artwick, Claudette G. 2013. "Reporters on Twitter." *Digital Journalism* 1 (2): 212–228 (doi: 10.1080/21670811.2012.744555).

Baldwin, Alan. 2012. "Inside the World of Olympic Timing: 'We Could Measure in Millionths of Seconds but That Wouldn't Be Fair.'" *The Independent*, July 25 (http://www.independent.ie/sport/other-sports/olympics/other-news/inside-the-world-of-olympic-timing-we-could-measure-in-millionths-of-seconds-but-that-wouldnt-be-fair-26879706.html).

Barassi, Veronica, and Emiliano Trere. 2012. "Does Web 3.0 Come After Web 2.0? Deconstructing Theoretical Assumptions through Practice." *New Media & Society* 14 (8): 1269–1285 (doi: 10.1177/1461444812445878).

Barkett, B. 2009. "Technologies for Monitoring Human Player Activity within a Competition." In *Digital Sport for Performance Enhancement and Competitive Evolution: Intelligent Gaming Technologies*, ed. N. K. Pope, K.-A. Kuhn, and J. J. Forster. Information Science Reference.

BBC. 2012. "Rural Chinese Get Online as Mobile Overtakes Desktop." BBC News, July 19 (http://www.bbc.com/news/technology-18900778).

BBC. 2014. "Video Games Should Be in Olympics, Says Warcraft Maker." BBC News (http://www.bbc.co.uk/news/technology-30597623).

BBC. 2015. "Is Computer Gaming Really Sport?" iWonder.

BBC. 2016. "The Beautiful Gamers." BBC Sport (http://www.bbc.co.uk/programmes/p03g8cyj).

Beckett, Charlie. 2008. *SuperMedia: Saving Journalism So It Can Save the World.* Wiley-Blackwell.

Beckett, Charlie, and James Ball. 2012. *WikiLeaks: News in the Networked Era.* Polity Press.

Beijing Organising Committee for the Olympic Games. 2007. Service Guide for Foreign Media Coverage of the Beijing Olympic Games and the Preparatory Period (http://bb.china-embassy.org/eng/xwfw/P020070904068474220087.pdf).

Bijker, Wiebe E. 1995. *Of Bicycles, Bakelites, and Bulbs: Toward a Theory of Sociotechnical Change.* MIT Press.

Blascovich, Jim. and Bailenson, Jeremy, 2011. *Infinite Reality: The Hidden Blueprint of our Virual Lives.* HarperCollins.

Borowy, Michael, and Dal Yong Jin, 2013. "Pioneering eSport: The Experience Economy and the Marketing of Early 1980s Arcade Gaming Contests." *International Journal of Communication* 7: 2254–2274.

Bostrom, N. 2003. "Are You Living in a Computer Simulation?" *Philosophical Quarterly* 53 (211): 243–255.

Bovill, M., and S. Livingstone. 2001. "Bedroom Culture and the Privatization of Media Use." In *Children and Their Changing Media Environment: A European Comparative Study*, ed. S. Livingstone and M. Bovill. Erlbaum.

Boyle, Raymond. 2006. *Sport Journalism: Context and Issues.* SAGE.

Boyle, Raymond, and Richard Haynes. 2004. *Football in the New Media Age.* Routledge.

British Olympic Association. 2007. *Team GB Advertising and Media Guidelines: A Guide for Olympians.* Officials, Sponsors, Media and Agents.

Burk, Dan. L. 2013. "Owning e-Sports: Proprietary Rights in Professional Computer Gaming." University of Pennsylvania Law Review 161 (6): 1535–1578 (http://scholarship.law.upenn.edu/cgi/viewcontent.cgi?article=1022&context=penn_law_review).

Byron, Tanya. 2008. Safe Children in a Digital World: The Report of the Byron Review. Department for Children, Schools and Families and Department for Culture, Media and Sport (http://webarchive.nationalarchives.gov.uk/20130401151715/http://www.education.gov.uk/publications/eOrderingDownload/DCSF-00334-2008.pdf).

Caddy, Becca. 2015. Twitter launches Moments to help users follow breaking news, October 7 (http://www.wired.co.uk/article/twitter-moments-breaking-news).

Carter, Chelsea J. 2012. "Viewers Outraged after NBC Cuts Away from Olympics Closing Ceremony." http://www.cnn.com/2012/08/13/sport/olympics-nbc-fail/

Castells, Manuel. 1996. *The Rise of the Network Society*. Blackwell.

Chambers, Deborah. 2012. "'Wii Play as a Family': The Rise in Family-Centred Video Gaming." *Leisure Studies* 31 (1): 69–82 (doi: 10.1080/02614367.2011.568065).

Chen, Chen-Yueh, Yi-Hsiu Lin, and Hui-Ting Chiu. 2013. "Development and Psychometric Evaluation of Sport Stadium Atmosphere Scale in Spectator Sport Events." *European Sport Management Quarterly* 13 (2): 200–215 (DOI: 10.1080/16184742.2012.759602).

China Internet Network Information Center. 2007. Statistical Survey Report on The Internet Development in China (http://www.apira.org/data/upload/pdf/Asia-Pacific/CNNIC/21streport-en.pdf).

Chowdbury, Saj. 2014. "France 3–Honduras 0." BBC Sport, June 15 (http://www.bbc.co.uk/sport/football/25285092).

Chozick, Amy. 2012. "NBC Unpacks Trove of Data From Olympics." *New York Times*, September 25 (http://www.nytimes.com/2012/09/26/business/media/nbc-unpacks-trove-of-viewer-data-from-london-olympics.html).

Clarke, Stephen, and Ross Stevenson. 1998. "The Games." Australian Broadcasting Corporation.

Cohen, Stanley. 1973. *Folk Devils and Moral Panics*. Paladin.

Collins, Harold Maurice. 2010. "The Philosophy of Umpiring and the Introduction of Decision-Aid Technology," *Journal of the Philosophy of Sport* 37: 135–146 (doi: 10.1080/00948705.2010.9714772).

Consalvo, Mia, and Konstantin Mitgutsch. 2013. *Sports VideoGames*. Routledge.

Corcoran, Liam. 2016. "The Biggest Facebook Publishers of January 2016." News-Whip (https://www.newswhip.com/blog/2016/02/the-biggest-facebook-publishers-of-january-2016).

Crawford, Garry. 2005. "Digital Gaming, Sport and Gender." *Leisure Studies* 24 (3): 259–270 (doi: 10.1080/0261436042000290317).

Crawford, Garry, and Victoria K. Gosling. 2009. "More Than a Game: Sports-Themed Video Games and Player Narratives." *Sociology of Sport Journal* 26 (1): 50–66.

Cross, Rod, and Graham Pollard. 2009. "Grand Slam Men's Singles Tennis 1991-2009 Serve Speeds and Other Related Data." *Itf Coaching & Sport Science Review* (http://www.physics.usyd.edu.au/~cross/GrandSlams.pdf#page=8).

Custer, C. 2014. "Why the World Cyber Games Got Cancelled: It's All Samsung's Fault." *Tech in Asia* (https://www.techinasia.com/why-the-world-cyber-games-got -cancelled-its-all-samsungs-fault).

Danzico, Matt. 2012. "Are Pro Video-Game Players Our 21st Century Athletes?" BBC News, August 28 (http://www.bbc.com/news/world-us-canada-19373708)

Dart, J. 2012. "New Media, Professional Sport and Political Economy." *Journal of Sport and Social Issues* 38 (6): 528–547 (doi: 10.1177/0193723512467356).

Duncan, J., M. Thorpe, and P. Fitzpatrick. 1996. "Sport on the Verge of the Third-Eye Era." *The Guardian*, [date to be added in proofs].

Edgar, Anthony. 2013. "Interaction with Anthony Edgar." *Sport in Society* 16 (9): 1206–1209 (doi: 10.1080/17430437.2013.819174).

Edgar, Anthony. 2014. "The Future of News and Sports Reporting at the Olympic Games and Other Major Events." Presented at Sportforum Schweiz, Lucerne.

Elias, Norbert and Eric Dunning. 1986. *Quest for Excitement: Sport and Leisure in the Civilizing Process*. Blackwell.

Elliott, Stuart. 2012. "With a Tattoo, Hanson Dodge Bets on Nick Symmonds." *New York Times*, July 4 (http://www.nytimes.com/2012/07/05/business/media/with-a -tattoo-hanson-dodge-bets-on-nick-symmonds.html).

Ellul, Jacques. 1964. *The Technological Society*. Vintage Books.

e-Sports Earnings. 2015. History of Earnings (http://www.esportsearnings.com/ history).

Finn, Jonathan. 2014. "Anatomy of a Dead Heat: Visual Evidence at the 2012 US Olympic Trials." *International Journal of the History of Sport* 31 (9): 976–993 (doi: 10.1080/09523367.2014.907795).

Flintham, M., R. Anastasi, S. Benford, A. Drozd, J. Mathrick, D. Rowland, A. Oldroyd, et al. 2003. "Uncle Roy All Around You: Mixing Games and Theatre on the City Streets." In *ACE'04 June 3–5, 2004, Singapore*, ed. M. Copier and J. Raessens. University of Utrecht and Digital Games Research Association.

4Creative. 2012. "Channel 4 Paralympics—Meet the Superhumans" (https://www .youtube.com/watch?v=tuAPPeRg3Nw).

Frank, Laura, and Betsy Cameron Williams. 2015. "Stats for London 2012 on Facebook. Facebook Newsroom (http://newsroom.fb.com/news/2012/08/stats-for -london-2012-on-facebook/).

Fromme, Johannes. 2003. "Computer Games as a Part of Children's Culture." *Game Studies* 3 (1) (http://www.gamestudies.org/0301/fromme).

Frost, Chris. 2011. *Journalism Ethics and Regulation*, third edition. Longman.

Fuller, Steve. 2013. *Humanity 2.0: What It Means to Be Human Past, Present and Future.* Palgrave Macmillan.

Funk, J. B. 2001. Children and Violent Video Games: Are There "High Risk" Players? (https://culturalpolicy.uchicago.edu/sites/culturalpolicy.uchicago.edu/files/funk1 .pdf).

Fusion Sport. 2012. "Talent ID" (http://fusionsport.com/portal/content/view/397/ 371/).

The Guardian. 2016. "Welcome to Your Cell." *The Guardian*, April 27 (http://www .theguardian.com/world/ng-interactive/2016/apr/27/6x9-a-virtual-experience -of-solitary-confinement).

Gaudiosi, John. 2015. "Why Coke Is Expanding beyond League of Legends eSports." http://fortune.com/2015/10/30/coke-expands-esports-reach/

Gelberg, J. Nadine. 1995. "The Lethal Weapon: How the Plastic Football Helmet Transformed the Game of Football, 1939–1994." *Bulletin of Science, Technology & Society* 15 (5–6): 302–309.

Gibson, William. 1984. *Neuromancer.* HarperCollins.

Gilchrist, Paul, and Neil Ravenscroft. 2013. "Space Hijacking and the Anarcho-Politics of Leisure." *Leisure Studies* 32 (1): 49–68 (doi: 10.1080/02614367.2012.680069).

Goel, Vindu. 2009. "Game Over: Scrabulous Shut Down on Facebook." *New York Times,* July 29 (http://bits.blogs.nytimes.com/2008/07/29/facebook-shuts-down -scrabulous/).

Goff, J. E., T. Asai, and S. Hong. 2014. "A Comparison of Jabulani and Brazuca Non-Spin Aerodynamics." *Proceedings of the Institution of Mechanical Engineers, Part P: Journal of Sports Engineering and Technology* 228 (3): 188–194 (doi: 10.1177/1754337114526173).

Griffiths, M. 2000. "Excessive Internet Use: Implications for Sexual Behaviour." *Cyberpsychology & Behavior* 3 (4): 537–552.

Grossman, Lev. 2009. "Iran's Protests: Why Twitter Is the Medium of the Movement." *Time,* June 19 (http://content.time.com/time/world/article/0,8599,1905125 ,00.html).

Guttmann, A. 1978. *From Ritual to Record: The Nature of Modern Sports.* Columbia University Press.

Hall, James. 2012. "Twitter Leads to under-Performance on Field of Play, Says Lord Coe." *The Telegraph,* December 30 (http://www.telegraph.co.uk/sport/olympics/ news/9771798/Twitter-leads-to-under-performance-on-field-of-play-says-Lord-Coe .html).

Hayles, N. Katherine. 1999. *How We Became Posthuman: Virtual Bodies in Cybernetics, Literature, and Informatics.* University of Chicago Press.

Heim, Michael. 1993. *The Metaphysics of Virtual Reality.* Oxford University Press.

Hemphill, D. A. 1995. "Revisioning Sport Spectatorism." *Journal of the Philosophy of Sport* 22: 48–60.

Hoberman, John. M. 1992. *Mortal Engines: The Science of Performance and the Dehumanization of Sport.* Free Press.

Howie, Luke, and Perri Campbell. 2015. "Fantasy Sports: Socialization and Gender Relations." *Journal of Sport and Social Issues* 39 (1): 61–77 (doi: 10.1177/0193723514533200).

House of Commons. 2005. London Olympics Bill (http://www.publications.parliament.uk/pa/cm200506/cmbills/045/2006045.pdf).

Huffman, R. K., and M. Hubbard. 1996. "A Motion Based Virtual Reality Training Simulator for Bobsled Drivers." In *The Engineering of Sport*, ed. S. Haake. Balkema.

Human Rights Watch. 2014. Letter to Sochi 2014 sponsors (https://www.hrw.org/sites/default/files/related_material/Olympics%20letter%20FINAL%20with%20logos.pdf).

Hutchins, Brett. 2008. Signs of Meta-Change in Second Modernity: The Growth of e-Sport and the World Cyber Games. *New Media & Society* 10 (6): 851–869.

Hutchins, Brett, David Rowe, and Andy Ruddock. 2009. "'It's Fantasy Football Made Real': Networked Media Sport, the Internet, and the Hybrid Reality of MyFootballClub. "*Sociology of Sport Journal* 26 (1): 89–106.

IBM. 2014. "Data Is a Game Changer" (https://www-07.ibm.com/innovation/au/ausopen/serve.html).

IGN. 2014. "ESPN President Says eSports Is Not a Sport." *IGN News*, September 8 (http://www.ign.com/articles/2014/09/08/espn-president-says-esports-is-not-a-sport).

International Olympic Committee. 2008. IOC Blogging Guidelines for Persons Accredited at the Games of the XXIX Olympiad, Beijing 2008. Lausanne: International Olympic Committee (http://www.olympiatoppen.no/ol/tidligereol/beijing_2008/utovere/reglement/media3178.media).

International Olympic Committee. 2010. Olympic Marketing Fact File (http://www.olympic.org/Documents/fact_file_2010.pdf).

International Olympic Committee. 2012. IOC Social Media, Blogging and Internet Guidelines for Participants and Other Accredited Persons at the London 2012 Olympic Games (http://www.olympic.org/documents/games_london_2012/ioc_social_media_blogging_and_internet_guidelines-london.pdf).

International Olympic Committee. 2014. Olympic Marketing Fact File (http://www
.olympic.org/Documents/IOC_Marketing/OLYMPIC_MARKETING_FACT_%20FILE
_2014.pdf).

International Olympic Committee. 2015a. The Olympic Charter (http://www
.olympic.org/Documents/olympic_charter_en.pdf).

International Olympic Committee. 2015b. IOC awards all TV and multiplatform
broadcast rights in Europe to Discovery and Eurosport for 2018-2024 Olympic Games
(http://www.olympic.org/news/ioc-awards-all-tv-and-multiplatform-broadcast
-rights-in-europe-to-discovery-and-eurosport-for-2018-2024-olympic-games/
246462).

International Olympic Committee. 2016. IOC Marketing Media Guide: Olympic
Games Rio 2016. Lausanne, IOC.

Internet World Stats. 2012. "167,335,675 Estimated Internet Users in Africa for
2012Q2" (www.internetworldstats.com).

Internet World Stats. 2015. "Internet Usage Statistics for Africa: Africa Internet Usage
and 2015 Population Stats" (http://www.internetworldstats.com/stats1.htm).

Ives, Jeffrey C., William F. Straub, and Greg A. Shelley. 2002. "Enhancing Athletic
Performance Using Digital Video in Consulting." *Journal of Applied Sport Psychology*
14 (3): 237–245.

Jankowski, N. W. 2006. "Creating Community with Media: History, Theories and
Scientific Investigations." In *The Handbook of New Media: Updated Student Edition*, ed.
L. A. Lievrouw and S. Livingstone. SAGE.

Kamal, Ankit. 2011. "Exergaming—New Age Gaming for Health, Rehabilitation and
Education." In *Advanced Computing* (*Communications in Computer and Information
Science*, volume 133) (doi: 10.1007/978-3-642-17881-8_40).

Kember, S. 1998. *Virtual Anxiety: Photography, New Technologies, and Subjectivity.*
Manchester University Press.

Kidd, Bruce. 2013. "The Olympic Movement and the Sports–Media Complex." *Sport
in Society* 16 (4): 439–448 (doi: 10.1080/17430437.2013.785754).

Kim, Yeojin, and Wilson Lowrey. 2014. "Who Are Citizen Journalists in the
Social Media Environment?" *Digital Journalism* 3 (2): 298–314 (doi: 10.1080/
21670811.2014.930245)

Kline, S. Dyer, Witherford, N. and De Peuter, G. 2003. *Digital Play: The Interaction of
Technology, Culture and Marketing.* McGill–Queen's University Press.

Kücklich, Julian. 2005. "Precarious Playbour: Modders and the Digital Games
Industry." *Fibreculture Journal*, issue 5 (http://five.fibreculturejournal.org/fcj-025
-precarious-playbour-modders-and-the-digital-games-industry/).

Kuhn, Kerri-Ann. 2009. "The Market Structure and Characteristics of Electronic Games." In *Digital Sport for Performance Enhancement and Competitive Evolution*, ed. N. K. Pope, K.-A. Kuhn, and J. J. Forster. Information Science Reference.

Lahajnar, Leon, Andrej Kos, and Bojan Nemec. 2008. "Skiing Robot—Design, Control, and Navigation in Unstructured Environment." *Robotica* 27 (4): 567–577 (doi: 10.1017/S0263574708004955).

Laird, Chris. 2012. "How Instagram Is Winning Gold at the 2012 Olympics." Mashable (http://mashable.com/2012/08/11/instagram-olympics-infographic/#XUZ1vw VqFZqM).

Lee, Hyuck-gi, and Won-hee Lee. 2012. "Presence in Virtual Golf Simulators : The Effects of Presence on Perceived Enjoyment, Perceived Value, and Behavioral Intention" (doi: 10.1177/1461444812464033).

Lee, Seunghwan, Won Jae Seo, and B. Christine Green. 2013. "Understanding Why People Play Fantasy Sport: Development of the Fantasy Sport Motivation Inventory (FanSMI)." *European Sport Management Quarterly* 13 (2): 166–199 (doi: 10.1080/16184742.2012.752855).

Lenskyj, H. 2002. *The Best Olympics Ever? Social Impacts of Sydney 2000*. State University of New York Press.

Lenskyj, Helen. 2006. "Alternative Media Versus Olympic Industry." In *Handbook of Sports and Media*, ed. A. A. Raney and J. Bryant. Routledge.

Lester, Cheryl. 2006. From Columbine to Red Lake: Tragic Provocations for Advocacy. *American Studies (Lawrence, Kan.)* 47 (1): 133–153.

Levy, Neil. 2012. "Your Brain on the Internet: A Response to Susan Greenfield." The Conversation (https://theconversation.com/your-brain-on-the-internet-a-response -to-susan-greenfield-8694).

Liebermann, Dario G., and Ian M. Franks. 2004. "The Use of feedback-based technologies." In *Notational Analysis of Sport*, second edition, ed. M. Hughes and I. M. Franks. Routledge.

Liebermann, Dario G., Larry Katz, Mike D. Hughes, Roger M. Bartlett, Jim McClements, and Ian M. Franks. 2002. "Advances in the Application of Information Technology to Sport Performance." *Journal of Sports Sciences* 20 (10): 755–769 (doi: 10.1080/026404102320675611).

Lin, J. 2014. "When Photos Come to Life: The Art of the Cinemagraph." http://time .com/3388024/when-photos-come-to-life-the-art-of-the-cinemagraph/

Livadas, P. 2005. Greek Olympic Know-How to Beijing. Greek Embassy. http://www .greekembassy.org/

Lomax, R. G. 2006. "Fantasy Sports: History, Game Types, and Research." In *Handbook of Sports and Media*, ed. A. A. Raney and J. Bryant. Routledge.

Love, Tom. 2011. *First Screening*, 88–91. SportsPro. [volume number or date to be inserted in proofs]

MacKinnon, R. 1997. "Virtual Rape." *Journal of Computer-Mediated Communication* 2 (4) (http://onlinelibrary.wiley.com/doi/10.1111/j.1083-6101.1997.tb00200.x/full).

Macutkiewicz, David, and Caroline Sunderland. 2011. "The Use of GPS to Evaluate Activity Profiles of Elite Women Hockey Players during Match-Play." *Journal of Sports Sciences* 29 (June): 967–973 (doi: 10.1080/02640414.2011.570774).

Mare, Admire. 2013. "A Complicated but Symbiotic Affair : The Relationship between Mainstream Media and Social Media in the Coverage of Social Protests in Southern Africa." *Ecquid Novi: African Journalism Studies* 34 (1): 83–98 (doi: 10.1080/02560054.2013.767426).

Malik, Shiv. 2012. "Twitter Suspends Account for Using London 2012 Olympics Logo." *The Guardian*, May 23 (http://www.theguardian.com/sport/2012/may/23/twitter-london-2012-olympic-logo).

Marik, Janica. 2013. "Gaming at the e-Sport Event: Mediatized Confrontations (Re) Negotiating Sport, Body and Media." In *Playing with Virtuality: Theories and Methods of Computer Game Studies*, ed. B. Bigl and S. Stoppe. Peter Lang.

Markham, Annette N. 1998. *Life Online: Researching Real Experience in Virtual Space*. AltaMira.

McArdle, D. 2000. "One Hundred Years of Servitude: Contractual Conflict in English Professional Football before Bosman." *Web Journal of Contemporary Legal Issues* 2 (http://webjcli.ncl.ac.uk/2000/issue2/mcardle2.html).

McGinnis, P. M. 2000. "Video technology for coaches." *Track Coach* 152 (summer): 4857–4862.

McGonigal, J. 2003. "'This Is Not a Game': Immersive Aesthetics and Collective Play." *Fineart Forum* 17 (8) (https://janemcgonigal.files.wordpress.com/2010/12/mcgonigal-jane-this-is-not-a-game.pdf).

McLuhan, Marshall. 1964. *Understanding Media: The Extensions of Man*. McGraw-Hill.

McMahon, D. 2014. "A Cyclist in the Tour De France Was Busted for Using His Cellphone at Nearly 40 MPH." Business Insider (http://uk.businessinsider.com/tour-de-france-cyclist-busted-using-cellphone-2014-7?r=US&IR=T).

Miah, Andy. 2002. "Immersion and Abstraction in Virtual Sport." In *Sport Technology: History, Philosophy and Policy*, ed. A. Miah and S. B. Eassom. Elsevier.

Miah, Andy. 2004. *Genetically Modified Athletes: Biomedical Ethics, Gene Doping and Sport*. Routledge.

Miah, Andy. 2011. "Towards Web 3.0: Mashing Up Work and Leisure." In *The New Politics of Leisure and Pleasure*, ed. P. Bramham and S. Wagg. Palgrave Macmillan.

Miah, Andy. 2014. "Tweeting the Olympic Games." In *The Routledge Handbook of the London 2012 Olympic*, ed. V. Girginov. Routledge.

Miah, A., and E. Rich. 2013. "The Body, Health and Illness." In *The Media: An Introduction*, third edition, ed. D. Albertazzi and P. Cobley. Routledge.

Millington, B. 2014. "Amusing Ourselves to Life : Fitness Consumerism and the Birth of Bio-Games." *Journal of Sport and Social Issues* 38 (6): 491–508 (doi: 10.1177/0193723512458932).

Mitchell, William. 1995. *City of Bits: Space, Place, and the Infobahn*. MIT Press.

Morgan, William. J. 1994. *Leftist Theories of Sport: A Critique and Reconstruction*, ed. B. G. Rader and R. Roberts. University of Illinois Press.

Morozov, Evgeny. 2009. "Iran Elections: A Twitter Revolution?" *Washington Post*, June 17 (http://www.washingtonpost.com/wp-dyn/content/discussion/2009/06/17/DI2009061702232.html).

Morris, B. S., and J. Nydahl. 1985. "Sports Spectacle as Drama: Image, Language and Technology." *Journal of Popular Culture* 18 (4): 101–110.

Morton, John. 2011. "Twenty Twelve." BBC (http://www.bbc.co.uk/programmes/b01f87nh).

Murray, John. Henry. 2000. Legislative Assembly, Wednesday 11 October (https://www.parliament.nsw.gov.au/prod/parlment/hanstrans.nsf/V3ByKey/LA20001011/$File/522la075.pdf).

NBC Sports Group. 2015. "Super Bowl XLIX on NBC Is Most-Watched Show in U.S. Television History." NBC Sports, February 2 (http://nbcsportsgrouppressbox.com/2015/02/02/super-bowl-xlix-on-nbc-is-most-watched-show-in-u-s-television-history/).

Neilson, Brett. 2002. "Bodies of Protest: Performing Citizenship at the 2000 Olympic Games." *Continuum (Perth)* 16 (1).

Newzoo. 2016. "The Global Games Market Reaches $99.6 Billion In 2016, Mobile Generating 37%." Newzoo, April 21 (https://newzoo.com/insights/articles/global-games-market-reaches-99-6-billion-2016-mobile-generating-37/).

Ovide, Shira. 2012. "Twitter Embraces Olympics to Train for the Big Time." *Wall Street Journal*, July 23 (http://www.wsj.com/articles/SB10000872396390444025204577543313839816248).

Owen, David. 2014. "Cuban Deal for Rio 2016 Takes IOC's Total Broadcast Income in Quadrennium to beyond $4 Billion." Inside The Games (http://www .insidethegames.biz/articles/1024469/cuban-deal-for-rio-2016-takes-ioc-s-total -broadcast-income-in-quadrennium-to-beyond-4-billion).

Pargman, D. 2000. "The Fabric of Virtual Reality—Courage, Rewards and Death in an Adventure MUD." *M/C* 3 (5) (http://journal.media-culture.org.au/0010/mud. php).

Peck, Tom. 2012. "Father of Olympic Branding: My Rules Are Being Abused." *The Independent*, July 20 (http://www.independent.co.uk/sport/olympics/news/father-of -olympic-branding-my-rules-are-being-abused-7962593.html).

Perl, Jürgen, and Daniel Memmert. 2011. "Net-Based Game Analysis by Means of the Software Tool SOCCER." *International Journal of Computer Science in Sport* 10: 77–84.

Peterson, Søren Mørk. 2007. "Mundane Cyborg Practice: Material Aspects of Broadband Internet Use. "*Convergence* 13 (1): 79–91.

Piccini, Angela. 2013. "Olympic City Screens: Media Matter, and Making Place." In *The Oxford Handbook of the Archaeology of the Contemporary World*, ed. P. Graves-Brown and R. Harrison. Oxford University Press.

Plant, Sadie. 2003. *On the Mobile: The Effects of Mobile Telephones on Social and Individual Life*. Motorola.

Plunkett, John. 2011. "BBC Denies Olympics Comedy Stole from Australian TV Show." *The Guardian*, March 16 (http://www.theguardian.com/media/2011/mar/16/ bbc-olympics-comedy-twenty-twelve).

Preuss, H. 2006. *The Economics of Staging the Olympics: A Comparison of the Games 1972–2008*. Edward Elgar.

Proteus. 2015. "U.S. FDA Accepts First Digital Medicine New Drug Application for Otsuka and Proteus Digital Health—Proteus Digital Health" (http://www.proteus .com/press-releases/u-s-fda-accepts-first-digital-medicine-new-drug-application-for -otsuka-and-proteus-digital-health/).

ProZone. 2014. "Big Data for a Bigger Picture" (http://prozonesports.com/big -data-bigger-picture/).

Ravensbergen, David. 2008. "Books Talking with Richard Florida." *The Tyee*, August 5 (http://thetyee.ca/Books/2008/08/05/Florida/).

Reed, Sada. 2013. "American Sports Writers' Social Media Use and Its Influence on Professionalism." *Journalism Practice* 7 (5): 555–571 (doi: 10.1080/ 17512786.2012.739325)>

Reisinger, Don. 2015. "China's Internet Users Soar as Mobile Drives Growth." *CNET*, February 3 (http://www.cnet.com/uk/news/chinas-internet-user-base-continues-to-soar-as-mobile-drives-growth/).

Rheingold, Howard. 1993. *The Virtual Community*. Addison-Wesley.

Rich, Emma. and John Evans. 2005. "'Fat Ethics'—the obesity discourse and body politics." *Social Theory & Health* 3 (4): 341–358.

Rich, Emma, and Andy Miah. 2009. Prosthetic Surveillance: The Medical Governance of Healthy Bodies in Cyberspace. *Surveillance & Society* 6 (2): 163–177.

Roberts, T. J. 1992. "The Making and Remaking of Sport Actions." *Journal of the Philosophy of Sport* 19: 15–29.

Robertson, Adi. 2013. "US Visa Bureau Says 'League of Legends' Is a Professional Sport." *The Verge*, July 13 (http://www.theverge.com/2013/7/13/4520188/us-citizenship-immigrations-to-give-league-of-legends-players-sports-visas).

Rogers, Katie. 2012. "<eref>, Olympic athletes take to Twitter to rally against strict sponsorship rules" *The Guardian*, July 31 (https://www.theguardian.com/sport/2012/jul/31/olympic-athletes-twitter-sponsorship-rules).

Ruggill, Judd Ethan, Ken S. McAllister, and David Menchaca. 2004. "The Gamework." *Communication and Critical/Cultural Studies* 1 (4): 297–312.

Rumsby, Ben. 2014. "Football Manager Player Database to Be Used by Premier League Clubs after Deal with Prozone Sports." *The Telegraph*, August 11 (http://www.telegraph.co.uk/sport/football/competitions/premier-league/11025724/Football-Manager-player-database-to-be-used-by-Premier-League-clubs-after-deal-with-Prozone-Sports.html).

Rushton, Bruce. 2013. "Backdooring It: Defense Maneuvers around Setback." *Illinois Times*, May 29 (http://illinoistimes.com/article-11440-backdooring-it.html).

Seeking Alpha. 2007. "Sohu.com Q1 2007 Earnings Call Transcript" (http://seekingalpha.com/article/34174-sohu-com-q1-2007-earnings-call-transcript).

Seo, Yuri, and Sang-Uk Jung. 2014. Beyond Solitary Play in Computer Games: The Social Practices of eSports. *Journal of Consumer Culture*. doi:10.1177/1469540514553711.

Shapiro, Rebecca. 2013. "NY Times Runs Instagram Photo on Front Page." Huffington Post, April 1 (http://www.huffingtonpost.com/2013/04/01/ny-times-instagram-photo-front-page_n_2991746.html).

Sherwood, M., and M. Nicholson. 2012. "Web 2.0 Platforms and the Work of Newspaper Sport Journalists." *Journalism* 14 (7): 942–959 (doi:10.1177/1464884912458662).

Silbermann, L. 2009. "Double Play: How Video Games Mediate Physical Perfor-mance for Elite Athletes." In *Digital Sport for Performance Enhancement and Competitive Evolution*, ed. N. K. Pope, K.-A. Kuhn, and J. J. Forster. Information Science Reference.

Smith, Briar. 2008. "Journalism and the Beijing Olympics: Liminality with Chinese Characteristics." In *Owning the Olympics: Narratives of the New China*, ed. M Price and D. Dayan. University. University of Michigan Press.

Smith, Marquard. 2005. *Stelarc: The Monograph*. MIT Press.

Smith, Murray. 1995. "Film Spectatorship and the Institution of Fiction." *Journal of Aesthetics and Art Criticism* 53 (2): 113–125.

Smith, Jacob. 2004. "I Can See Tomorrow in Your Dance: A Study of Dance Dance Revolution and Music Video Games." *Journal of Popular Music Studies* 16 (1): 58–84.

Sorrentino, Ruth Morey, Richard Levy, Larry Katz, and Xiufeng Peng. 2005. "Virtual Visualization: Preparation for the Olympic Games Long-Track Speed Skating." *International Journal of Performance Analysis in Sport* 4 (_): 40–45 (http://www.ucalgary.ca/~rmlevy/Publications/Sorrentino_Levy_Katz_Peng_IJCSS-Volume4_1_2005.pdf).

Souppouris, A. 2014. "Virtual Reality Made Me Believe I Was Someone Else." *The Verge* (http://www.theverge.com/2014/3/24/5526694/virtual-reality-made-me-believe-i-was-someone-else)

Sponsorship Intelligence. 2008. "Games of the XXIX Olympiad, Beijing 2008: Global Television and Online Media Report" (http://www.olympic.org/Documents/IOC_Marketing/Broadcasting/Beijing_2008_Global_Broadcast_Overview.pdf).

Sponsorship Intelligence. 2012. "London 2012 Olympic Games: Global Broadcast Report" (http://www.olympic.org/Documents/IOC_Marketing/Broadcasting/London_2012_Global_Broadcast_Report.pdf).

Squatrigilia, Chuck. 2012. "Olympic Runner Auctions Ad Space—on His Body." *Wired*, January 9 (http://www.wired.co.uk/news/archive/2012-01/09/runner-auctions-ad-space).

Sterling, Bruce. 1997. "Unstable Networks." In *Digital Delirium*, ed. A. Kroker and M. Kroker. St. Martin's Press.

Stivers, R. 2001. *Technology as Magic: The Triumph of the Irrational*. Continuum.

Stoddart, B. 1997. "Convergence: Sport on the Information Superhighway." *Journal of Sport and Social Issues* 21 (1): 93–102.

Streberny, A., and G. Khiabany. 2010. *Blogistan: The Internet and Politics in Iran*. I.B. Tauris.

Stuart, Keith. 2014, Aug 12. "Why Clubs Are Using Football Manager as a Real-Life Scouting Tool." http://www.theguardian.com/technology/2014/aug/12/why-clubs-football-manager-scouting-tool

Suits, Bernard. 1967. "What Is a Game?" *Philosophy of Science* 34 (2): 148–156 (http://www.simulationtrainingsystems.com/game/).

Suits, B. 1978. *The Grasshopper: Games, Life, and Utopia*. University of Toronto Press.

SuperData. 2015. "eSports Market Brief 2015/2016 Update" (http://superdata-research.myshopify.com/products/esports-market-brief-2015

Supponor. 2015. "How It Works" http://www.supponor.com/dbr-live/how-it-works

Sweney, M. 2015. "Facebook Instant Articles: BBC News and *Guardian* Sign Up to Initiative." *The Guardian*, May 13 (http://www.theguardian.com/media/2015/may/13/bbc-news-guardian-facebook-instant-articles).

Taylor, T. L. 2012. *Raising the Stakes: E-Sports and the Professionalization of Computer Gaming*. MIT Press.

Tenner, E. 1996. *Why Things Bite Back: Predicting the Problems of Progress*. Fourth Estate.

Thompson, Derek. 2014. "Half of Broadcast TV Viewers Are 54 and Older—Yikes." *The Atlantic*, March 5 (http://www.theatlantic.com/business/archive/2014/03/half-of-broadcast-tv-viewers-are-54-and-older-yikes/284256/).

Toffler, Alan. 1970. *Future Shock*. Pan Books.

Toney, James. 2012. *Sports Journalism: The Inside Track*. Bloomsbury Sport.

Trabal, Patrick. 2008. "Resistance to Technological Innovation in Elite Sport." *International Review for the Sociology of Sport* 43 (3): 313–330 (doi: 10.1177/1012690208098255).

Turkle, Sherry. 1995. *Life on the Screen: Identity in the Age of the Internet*. Weidenfeld and Nicolson.

Turkle, Sherry. 2005. *The Second Self: Computers and the Human Spirit*, twentieth anniversary edition. MIT Press.

Turkle, Sherry. 2011. *Alone Together: Why We Expect More from Technology and Less from Each Other*. Basic Book.

Turner, Mark. 2013. "Modern 'Live' Football: Moving from the Panoptican Gaze to the Performative, Virtual and Carnivalesque." *Sport in Society* 16 (1): 85–93 (doi: 10.1080/17430437.2012.690404).

Vastag, B. 2004. "Does Video Game Violence Sow Aggression?" *Journal of the American Medical Association* 291 (15): 1822–1824 (doi: 10.1001/jama.291.15.1822).

Walser, R. 1991. "Elements of a Cyberspace Playhouse." In *Virtual Reality: Theory, Practice and Promise*, ed. S. K. Helsel and J. P. Roth. Meckler.

Weber, Ian. 2005. "Digitizing the Dragon: Challenges Facing China's Broadcasting Industry." *New Media & Society* 7 (6): 791–809.

Weitao, Li. 2007. "Internet Users to Log in at World No. 1." *China Daily*, January 4 (http://www.chinadaily.com.cn/china/2007-01/24/content_790804.htm).

West Ham United Football Club . 2016. "Hammers sign E-Sports star." *West Ham United*, May 6 (http://www.whufc.com/News/Articles/2016/May/6-May/Enter-the -Dragonn).

Wikipedia. 2015. "List of best-selling video game franchises" (https://en.wikipedia .org/wiki/List_of_best-selling_video_game_franchises).

Wolf, Gary. 2010. "The Data-Driven Life." *New York Times*, April 28 (http://www .nytimes.com/2010/05/02/magazine/02self-measurement-t.html?_r=0).

Xiong, Qu. 2008. "Internet Rights Holders for Beijing 2008 Olympic Games" (http:// www.cctv.com/english/20080806/106217.shtml).

Index